A critical approach to the social acceptance of renewable energy infrastructures

Susana Batel • David Rudolph
Editors

A critical approach to the social acceptance of renewable energy infrastructures

Going beyond green growth and sustainability

palgrave
macmillan

Editors
Susana Batel
Instituto Universitário de Lisboa
(ISCTE-IUL), Cis-IUL
Lisboa, Portugal

David Rudolph
Department of Wind Energy
Technical University of Denmark
Roskilde, Denmark

ISBN 978-3-030-73698-9 ISBN 978-3-030-73699-6 (eBook)
https://doi.org/10.1007/978-3-030-73699-6

This Palgrave Macmillan imprint is published by the registered company Springer Nature Switzerland AG.
The registered company address is: Gewerbestrasse 11, 6330 Cham, Switzerland

Acknowledgements

The idea for this book evolved as a result of many wanderings, encounters and experiences of each one of us individually along our personal and academic pathways, but also as a result of some conversations shared between us and other colleagues as we met in the context of specific networks and related activities. Examples of those are one meeting of the Task 28—Social Acceptance of Wind of the International Energy Agency that took place in Roskilde, Denmark, in 2018, as well as activities related with MISTRAL (Multi-sectoral approaches to innovative skills training for renewable energy and social acceptance), a Horizon 2020 Marie Sklodowska Curie Actions (MSCA)-funded Innovative Training Network, which served as meeting points for reflecting about many of the issues that this book addresses and proposes. We would like to thank these networks and the colleagues therein involved for providing interesting and stimulating spaces for discussion; they surely inspired and shaped our own discussions and the content of this book.

Susana Batel would also like to thank the Portuguese Science Foundation for supporting her work through Portuguese national funds through in the context of Norma Transitória—DL57/2016/CP1359/CT0039.

David Rudolph would like to gratefully acknowledge support from the International Energy Agency (IEA) Task 28—Social Acceptance.

We would like to thank all authors for their contributions in the critical spirit of this book and for addressing our various requests in accordance with required formalities and specifications, even though our positionalities and subjectivations as critical scholars necessarily presuppose a sceptical or even disapproving stance towards certain tasks and compliances, and despite the challenging conditions we all faced during the writing process throughout 2020. Our biggest thanks goes therefore out to all the authors as well as staff at Palgrave, in particular Rachael Ballard, Joanna O'Neill and Jasper Asir, who made this book possible and as inspiring and relevant to supporting fairer and more just energy transitions and societies, as we believe it is.

November 2020 Lisbon and Roskilde

Contents

Notes on Contributors

Susana Batel is an integrated researcher at the Center for Psychological Research and Social Intervention (Cis) of the University Institute of Lisbon (Iscte), Portugal. Her research adopts a critical and interdisciplinary perspective to examine the relationship between people, the territory and the climate crisis, specifically around energy transitions towards carbon neutrality, and associated social justice and political participation issues. She is also interested in how discourse and communication relate with social change and resistance, and has published in journals like *Antipode, Local Environment: the International Journal of Justice and Sustainability, Energy Research & Social Science, and Journal of Environmental Psychology.*

Ellen Chappell completed a Master of Environmental Studies at Dalhousie University. Her research focused on social factors influencing public perceptions and acceptance of wind energy. She has also assisted with research on factors influencing support of/opposition to different energy sources in Canada, with a focus on perceptions of coal, public acceptance of adaptation strategies for sea level rise in coastal communities and use of adaptive grazing strategies by cattle farmers. Ellen is working for the Canadian government in the field of environmental health and is interested in strategies to improve both environmental and human health.

Laura Tolnov Clausen is an associate professor at the Department of Global Development and Planning at the University of Agder, Norway. She holds a PhD from Roskilde University exploring democratic issues of environmental planning in Denmark. Her research interests include social, spatial and democratic dimensions of environment and sustainability issues, including work on energy democracy, community energy, public participation, place-identity, environmental justice and renewable energy. In recent years, her research has focused on public perceptions and community reactions to the siting of wind farms. Recent research projects explore public participation in wind power planning in Norway, Scotland and Denmark.

Eefje Cuppen is Professor of Governance of Sustainability at the Institute of Public Administration of the Faculty of Governance & Global Affairs, Leiden University, where she is developing the university's interdisciplinary research programme 'Liveable Planet'. Her research focuses on stakeholder participation and public engagement in sustainability transitions. She holds an MSc degree in Innovation Sciences from Eindhoven University of Technology and a PhD in Policy Sciences from VU University. She has published widely on design and methodology of stakeholder dialogue, public engagement and controversy in energy planning, responsible innovation and governance of industrial symbiosis, and is editor of *Sustainability Science*.

Patrick Devine-Wright is Professor in Human Geography at the University of Exeter. He is an environmental social scientist with an interest in community engagement and place attachment applied to the understanding of energy transitions and climate change. With expertise spanning human geography and environmental psychology, he was cited in the top 1% of global scholars in 2019 (Web of Science). He is an Intergovernmental Panel on Climate Change (IPCC) lead author and Chair of the Devon Net Zero Task Force. He received a Distinguished Visiting Scientist award from Commonwealth Scientific and Industrial Research Organisation (CSIRO), Australia (2012–2013), and is a board member of various journals, including *Global Environmental Change*, *Energy Research and Social Science* and the *Journal of Environment Psychology*.

Andrés Di Masso, PhD is professor at the Social Psychology Section and coordinator of the Research Group in Interaction and Social Change (GRICS-2017AGAUR-1500), University of Barcelona (Spain). His research focuses on the everyday politics of people-place relations, applied to varied issues such as gentrification, xeno-racism, place attachment, postcolonial geographies and new models of fatherhood.

Alexander Dunlap is a postdoctoral research fellow at the Centre for Development and the Environment, University of Oslo. His work has critically examined police-military transformations, market-based conservation, wind energy development and extractive projects more generally in both Latin America and Europe. He is the author of two books: *Renewing Destruction: Wind Energy Development, Conflict and Resistance in a Latin American Context* (2019) and (with J. Jakobsen) *The Violent Technologies of Extraction* (2020, Palgrave). He has also published in *Anarchist Studies*, the *Journal of Peasant Studies*, *Political Geography*, the *Journal of Political Ecology*, *Globalizations*, the *Journal of Genocide Research* and more.

Aiste Garneviciene holds an MA in Anthropology from the University of Aberdeen. In 2019, she completed her MSc in Energy, Society and Sustainability from The University of Edinburgh. Her dissertation explored injustices associated with the development of utility-scale solar in Malawi and analysed the role sociotechnical imaginaries play in the support and acceptance of renewable energy projects. Aiste works for the UK gas and electricity market regulator (Ofgem), informing the delivery of social and environmental schemes.

Mike Gismondi is an adult educator and professor emeritus with Athabasca University, Canada's Open University. He is researching the role of social and solidarity economy practitioners in sustainability transitions, community energy in Alberta and how peer-to-peer learning might accelerate systems change. Mike is co-editor of *Scaling Up: The Convergence of Social Economy and Sustainability* (2016). He is working with colleagues in the Synergia Cooperative Institute to re-offer *Towards Co-operative Commonwealth: Transition in a Perilous Age*, a popular education Massive Open Online Courses (MOOC) designed to share charismatic projects, their practice and strategies to advance low-carbon, just futures.

Rebecca Grant is an Economic and Social Research Council-funded PhD candidate at The University of Edinburgh. Her PhD examines energy (in)justices associated with the promotion and use of off-grid solar photovoltaic systems in rural Rwanda. This draws on a mixture of methods, including interviews, participatory mapping, document analysis and geospatial analysis. She completed her MRes at the University of St Andrews in 2019, where she examined energy injustices associated with off-grid hydroelectricity in Malawi.

Claire Haggett is a senior lecturer at the School of Social and Political Science, The University of Edinburgh. For the last 20 years, she has researched public responses to renewable energy developments, focusing on engagement processes and benefits, community energy schemes and the role of social and environmental assessment. Claire has recently led a series of research projects for the Scottish government on community engagement and renewables, is the author of a number of widely cited publications on these topics and has conducted research funded by the Economic and Social Research Council (ESRC), Natural Environment Research Council (NERC), the UK Energy Research Centre and the EU.

Lorelei Hanson is Associate Professor of Environmental Studies at Athabasca University. Her research interests include energy transition, critical sustainability, environmental history and environmental advocacy. An active community-based environmental researcher, Lorelei is collaborating on an Indigenous-led counter-mapping project of Alberta's Bighorn Country, an Indigenous community-based environmental monitoring educational project, and undertaking research on community energy in Alberta. She is editor of *Public Deliberation on Climate Change: Lessons from Alberta Climate Dialogue* (2018) and has published in journals such as *Environmental Politics*, the *International Journal of Interdisciplinary Environmental Studies* and *Local Environment the International Journal of Justice and Sustainability*.

Julia Kirch Kirkegaard is a Science & Technology Studies (STS) researcher at DTU Wind Energy. She has published and taught on socio-technical controversies over the transition to renewable energy. In her research, she has focused empirically on the wind power sector—in particular in Danish, Chinese and American contexts—inquiring into the role of technological, scientific and economic expertise and innovation in the creation of electricity markets, together with the development of wind farms and the interplay with public engagement.

Olivier Labussière is a researcher at the French National Scientific Research Centre (CNRS). He is a member of the Pacte Social Sciences Research Centre (Grenoble, France), where he leads the interdisciplinary 'Environments' research group. He studies the relationships between energy, society and space, and their emerging environmental issues in different contexts (onshore, offshore, underground).

Rafaella Lenoir-Improta holds a PhD in Social and Environmental Psychology from the University of Barcelona. Her main research interest focuses on a critical view of people-place relations, applied to diverse topics as social acceptance of renewable energy, environmental justice, green infrastructures, place attachment, postcolonial studies, social inequality, gentrification, social inclusion and marginalized social groups.

Adolfo Mejía Montero has a bachelor's in Physics from the Autonomous National University of Mexico (UNAM), and an MSc in Sustainable Energy Systems from the School of Engineering, The University of Edinburgh. He recently completed his PhD from The University of Edinburgh, studying the development of utility-scale wind power in Oaxaca, Mexico through the conceptual lenses of Energy Justice. Adolfo is a course organizer and tutor at The University of Edinburgh, and a consultant for interdisciplinary energy projects in Mexico and the UK.

Alain Nadaï is a senior interdisciplinary social scientist at CIRED (International Research Center on Environment and Development), which is part of French CNRS. His research activity has been centred on environmental controversies and policies, landscape policies and energy transition policies. His research activity focuses on the sociotechnical, spatial and temporal changes induced by energy change processes. He has contributed as a leading author to the IPCC Special Report on Renewable Energy Sources and Climate Change Mitigation (SRREN, 2011), and co-edited a book about the democratic dimension of energy transition processes: *Energy Transitions—A Sociotechnical Inquiry* (Palgrave, 2018).

Sophie Nyborg is a Science & Technology Studies (STS) scholar with an interdisciplinary background in the natural and social sciences as well as in science journalism. She is currently a Researcher in the Department of Wind Energy at the Technical University of Denmark. Her research revolves around responsible research and innovation, and she publishes and teaches on public participation in techno-science, socio-technical controversies, co-creation, stakeholder engagement and user oriented design and innovation. Empirically, her research focuses mainly on the sustainable transition agenda and renewable and smart energy.

John Parkins is Professor of Sociology at the Department of Resource Economics and Environmental Sociology, University of Alberta. His research and teaching address the social context of resource development, renewable and community energy, public deliberation and environmental politics, and sustainable agriculture in Alberta. Recent publications include case studies of community energy in western Canada, barriers and opportunities for wind power development in Alberta, the social context of public engagement in fracking and forestry in Canada.

Udo Pesch is an associate professor at Delft University of Technology. He is affiliated to the Department of Values, Technology and Innovation, Faculty of Technology, Policy and Management. His disciplinary interests include responsible innovation, participatory decision-making, environmental politics, public policy and ethics. Recently published work includes articles on energy justice, sustainable innovation and moral emotions. He has published a wide range of articles and book chapters, among others, in *Social Studies of Science* (SSS), *Science, Technology & Human Values* (STHV), the *Journal of Responsible Innovation* (JRI), *Science and Engineering Ethics* (SEE) and *Technological Forecasting and Social Change* (TFSC).

David Rudolph is a researcher at the Department of Wind Energy, Technical University of Denmark. He holds a PhD in Human Geography from The University of Edinburgh, and his research seeks a critical geographical understanding of energy transitions. His work is concerned with contestations over onshore and offshore renewables, people-place relations, community participation, governmentality and radical planning, with a specific focus on social justice and rural areas. He has published and co-edited special theme issues on these topics in *Antipode, Environment and Planning C, Energy Research & Social Science, Energy Policy, Scottish Geographical Journal* and *Rural Studies.*

Kate Sherren is Professor of Landscape Social Science at the School for Resource and Environmental Studies, Dalhousie University. She researches the human dimensions of sustainability landscape transitions in diverse contexts, including agriculture, energy, cities and coasts. Her recent work focuses on the development and testing of a 'climax thinking' framework to understand and reduce resistance to climate-motivated actions such as nature-based coastal adaptation and renewable energy expansion, landscape culturomics using big data to improve social impact assessment, and cultural ecosystem services to inform landscape decision-making.

Dan van der Horst holds a personal Chair in Energy, Environment & Society at the School of Geosciences, The University of Edinburgh. He is teaching and learning about societal transitions towards sustainability. Dan is committed to collaborative, transdisciplinary, citizen-oriented and policy-facing research to help accelerate the adoption of low-carbon technologies and lifestyles, as well as to reducing poverty and inequalities. His publications range from sustainable land use policy, ecosystem services and nature-based solutions in anthropogenic landscapes to public engagement with renewable energy, energy literacy in the home and local responsibility 'to do our bit' in the face of global environmental crises.

List of Figures

PART I

Introduction

CHAPTER 1

A Critical Approach to the Social Acceptance of Renewable Energy Infrastructures

Susana Batel and David Rudolph

Introduction

Calls for increasing environmental sustainability, decarbonising the energy system, fighting climate change and related efforts are more and more pervasive in current societies and have become more visible in conventions and related treaties, policy initiatives and activists' actions—like the Paris Agreement, the European Green Deal and Fridays for Future protests, to name but a few. These highlight that ideas around fostering greener societies and promoting sustainability have already become quite consensual—notwithstanding the recent rise of right-wing, populist and

S. Batel (✉)
Instituto Universitário de Lisboa (ISCTE-IUL), Cis-IUL, Lisboa, Portugal
e-mail: susana.batel@iscte-iul.pt

D. Rudolph
Department of Wind Energy, Technical University of Denmark, Roskilde, Denmark
e-mail: dpru@dtu.dk

S. Batel, D. Rudolph (eds.), *A critical approach to the social acceptance of renewable energy infrastructures*,
https://doi.org/10.1007/978-3-030-73699-6_1

climate change denialist positions by certain groups, which are quite expressive in certain parts of the world, such as in the US or in Brazil, in reaction to a more widespread consensus over the need to make our societies more sustainable.

This consensus has also shaped—and often departed from—academic arenas, with the expanding research on the social acceptance of renewable energy innovation and associated technologies (RET) as a clear example of that. This area of research has massively grown in the past decade, mainly based on the attempt to understand the supposed paradox that despite the consensus over the importance of tackling climate change and thus generating renewable energy as opposed to a continued utilisation of fossil fuels, public opposition to the deployment of RET is often found. The understanding of this paradox and consolidation of this area of research have had as a cornerstone the so-called triangle of social acceptance of renewable energy innovation, proposed to be constituted by interrelated realms of socio-political acceptance, market acceptance and community acceptance (Wüstenhagen et al., 2007). This area of research has thus been predominantly concerned with a disentanglement of the drivers for and barriers to social (community, socio-political, market) acceptance, often with a view to overcoming barriers and fostering social acceptance, as already extensively discussed elsewhere (e.g. Aitken, 2010; Batel et al., 2013).

However, while doing so, it has pursued a normative frame that assumes that increased renewable energy generation per se, independently of how it is produced and what, where, by whom and for whom, can achieve increased environmental sustainability and tackle climate change. As such, it uncritically assumes that more just and sustainable societies can be achieved within the very same socio-economic and political structures that produce, use and capitalise fossil fuels for economic growth (Raman, 2013; also Howe & Boyer, 2016; Batel & Devine-Wright, 2017; Dunlap, 2018). This is founded in the fundamental underlying rationale of sustainable development to decouple development and growth from the use of fossil fuels and its disastrous consequences by replacing them with renewable resources. Hence, the decoupling is predominantly based on the substitution of fuels and a replacement of technologies without questioning inherent growth principles, but rather relying on market logics to facilitate and enable this transformation. Yet, the scope of the proposed solutions is therefore constrained within a business-as-usual approach concerning the political-economic premises driven by compound economic growth and technological interventions (Giotitsas et al., 2020). This institutional status quo also characterises a post-political condition in which techno-managerial planning replaces dissensual contestations (Swyngedouw, 2010) and where potential solutions are only earmarked and addressed in a highly selective manner. Hence, the underlying technocratic rationale problematises the fuel or technology as the ultimate issue to be solved, rather than the underlying processes that determine and aggravate the problem. Conversely, it also points at people and social framings as the associated obstacle to innovation and to the introduction of new technologies, as initially framed within the social acceptance literature primarily aimed at overcoming resistance and opposition (Batel, 2020). This technocratic approach of decoupling continuous growth from finite and adverse fossil fuel resources has also been denominated as a socio-ecological fix (McCarthy, 2015). The substitution of fossil fuels with renewable resources is supposed to provide a fix for interrelated environmental and economic crises of fossil-fuel-driven capitalism, as a 'way to renew accumulation on a more socially and environmentally sound basis' (McCarthy, 2015, p. 2491) by commodifying renewable resources. Green growth, sustainable development and ecological modernisation are founded on the economic strategy to appropriate and internalise environmental problems into new accumulation processes, thus co-opting a threat as an opportunity for new profitability and therefore maintaining the political-economic status quo (Kenis & Lievens, 2015, 2016). While turning a crisis into an opportunity for technological renewal through new accumulation cycles is considered to be a central prerequisite for environmental innovation (Bosch & Schmidt, 2019), some scholars (e.g. Brand & Wissen, 2018; Harris, 2011; McCarthy, 2015) question the capability of a renewed technocratic mastery of nature to bring about lasting environmental and social sustainability. This critique argues that a dominant focus on technological—or even socio-technical—solutions tends to externalise adverse social and environmental consequences still inherent in the capitalist social order and thus merely shifts disparities and uneven development, instead of fostering deep socio-ecological transformations. Hence, energy transitions comprise more than the replacement of fossil fuels and related technologies with renewable infrastructures and related socio-technical arrangements, and this is what social acceptance research is required to capture as well if it is to do justice to understanding the political-economic context of RET and to look beyond individualised and localised hindrances of technological advancements and transformations.

However, in relation to the socio-economic underpinnings, emerging ideals of energy democracy (Becker & Naumann, 2017) generally suggest that renewable energy as distributive forms of energy are less prone to capitalist imperatives of profit, private property and market logics, but allow for greater democratic ownership, community empowerment and distributive control (Daggett, 2020). Thus, the emergence of renewable energies has altered the conditions for social contestations and paved the way for new democratic opportunities. This facet has been particularly emphasised with regard to the possibilities of what has been termed as community energy (Creamer et al., 2019), which has evolved into an important strand of social acceptance literature, with locally anchored community energy projects generally deemed to be more just and acceptable. However, Eadson and Foden (2019) and Creamer et al. (2018) demonstrated that the enablement of decentralised community energy is nonetheless entangled with the ubiquity of market logics and institutional arrangements that do not operate outside dominant economic relations (see also van Veelen, 2020). In turn, community ownership of renewable energy can also be considered to serve as a 'fix' to cushion the consequences of neoliberal capitalism, in particular the effects of austerity politics in the UK (Lacey-Barnacle, 2020; Slee, 2015). Even more so, the notion of asset-based community development through harnessing local energy resources along with greater control over economic activities as a solution to local problems and the withdrawal of the state in providing central services may rather conceal a distortion of neoliberal values of privatisation and individual responsibilisation as empowerment (Clausen & Rudolph, 2020; Macleod & Emejulu, 2014). On the other hand, unless protective measures are in place in order to prevent a displacement of community energy through economies of scale, the utilisation of community or cooperatively organised renewable energy projects also tends to represent a niche or window of opportunity preceding the capitalisation of the renewable energy sector, as seen in Denmark (Gorroño-Albizu et al., 2019; Strachan et al., 2015). The capitalisation of renewable energy in frontrunner countries, such as Denmark and Germany, has been characterised not only by an inevitable takeover of RET developments through large commercial players, but also by technological advancements, market arrangements and shifted power relations, which impede an equal participation of less financially potent actors (Kirkegaard et al., 2020). This is particularly evident in the recent shift in the support system for renewable energies from fixed feed-in tariffs to more market-based instruments, like auctions, that tend to favour economies of scale. The large-scale development of renewable energies and its increased or existent detachment from the communities and local areas in which they are deployed is regarded as an essential constituent of local responses, which has resulted in extensive research dealing with community engagement in planning and the provision of community benefits (Batel, 2020; Baxter et al., 2020). Both aspects are associated with a reattachment of communities to renewable energy developments but give striking evidence of academic research being cleaved to dominant rationales previously employed to legitimate fossil-fuel-based growth and other extractive industries. The topic of community participation has been largely framed by an understanding of meaningful approaches to engagement, fair decision-making and the establishment of trust in authorities and developers as inevitable conditions for achieving legitimacy of private-sector-led developments (Aitken et al., 2016; Gross, 2007; Natarajan et al., 2018; Stober et al., 2021). Furthermore, the very idea of community benefits as a voluntary or mandatory distribution of profits and sharing of revenues from renewables (Cowell et al., 2011; Jørgensen, 2020; Rudolph et al., 2018) is indicative of the inability of the socio-economic principles guiding the deployment of renewables to automatically contribute to more just and equitable outcomes and to easily translate into more sustainable societies. Thus, certain research strands have been influenced by the socio-economic conditions in which the development of RET occurs and operates. Although recent social acceptance research has been increasingly spurred by justice-driven rationales with the aim to create more equitable outcomes (Batel, 2020), the underlying need to do so has so far remained within the scope of discussing instrumental approaches (e.g. improved procedures and community benefits) to cope with socio-economic conditions, rather than fundamentally scrutinising the overarching political-economic system. In addition, RET have evolved to critical infrastructures in many countries whose deployment has become increasingly exposed to and influenced by neoliberal rationales of development. This stage does no longer differentiate (novel) renewables from any other infrastructure developments but demands a critical approach to social acceptance that pays more attention and is more sensitive to the socio-economic and political dimensions of RET expansion, and their spatial, community and psycho-social associations and consequences. Hence, and as recently highlighted by Cara Daggett (2020), while energy research and social science scholarship has importantly brought to the fore the social and political dimensions of energy transitions, research is still often performed through a narrow focus on how fuels and technologies should be changed, which reflects a dominant, fossil-fuel-based narrative of growth, progress, expansion and innovation.

This book departs from this state of the art of research on the social acceptance of RET and of research on energy transitions more generally and argues for the need to go beyond green growth and sustainability discourses as the solutions and goals for the ecological crisis and to foster instead more critical understandings of what renewable energy is, how it relates with, reproduces and/or contests current neoliberal and capitalist systems, and the extent to which it can address the ecological crisis and associated social inequalities and injustices. It aims to identify and discuss specific ways through which research on the social acceptance of RET can or should overcome such depoliticised green growth and sustainability discourses and associated post-political configurations of governance and related assumptions and consequences (e.g. Batel et al., 2016; Lennon et al., 2020; Pepermans & Maeseele, 2014; Rudolph & Kirkegaard, 2018; Swyngedouw, 2011). As such, being critical, as here proposed, is to reveal ways in which we can actually identify, examine, and contest the power relations and associated inequalities created by the deployment of RET within the logics of green growth and depoliticised sustainability. This follows the tradition of critical research through a greater awareness of the social structures, power relations and material practices in order to understand and enable ways that promote equally the well-being of ecosystems, human and non-human beings. A critical approach does not just neutrally portray societal conditions and developments, but strives to be more normative in intervening in existing power relations and socio-political contestations. In doing so, the book addresses previous shortcomings and takes a critical stance towards previously unquestioned premises in order to provide a reflexive examination of how the social, historical, material and structural conditions variously shape the issue of the social acceptance of RET, rather than simply following technocratic principles and providing regulatory solutions to renewable energy and delivering the accompanying social acceptance research. In that regard, this book goes beyond previous proposals in this area that can be more situated within a criticism approach to research on the social acceptance of RET (Batel, 2020; also Nadaï and Labussière, this book), such as those viewing RET as leading to sustainable economic development (Warren et al., 2012); those framing it through the energy trilemma (Bridge et al.,

2018); or those arguing for more, better and varied participation in RET decision-making and energy transition (see Devine-Wright, 2011; Fraune et al., 2019; Holstenkamp & Radtke, 2018) or for more actor or individual-centred perspectives to better understand responses to RET (Upham et al., 2019), but largely without considering how participation and actors' responses are entangled with the larger socio-economic and political system. To acknowledge, examine and incorporate this entanglement of RET with the workings of neoliberal capitalism in research and practice is crucial. Only in this way can systematic and articulated efforts be made in line with fostering the required radical transformation that could arguably still impede the ecocide/genocide towards which humanity (in its human, non-human and ecological forms) is heading (Dunlap, 2020; Haraway, 2015). This demands us to adopt more holistic lenses of analysis (Brand & Wissen, 2018) that consider renewable energy in its most common forms as just another manifestation of the drive towards economic growth, commodification of nature and increasing individualisation, and associated global inequalities, local injustices and psycho-social impacts.

Thus, the book aims to delineate a more critical and politicised agenda for social acceptance research by focusing on what we consider to be five main axes for developing such a critical approach (see also Batel & Adams, 2016): (1) overcoming individualisation and socio-cognitive reductionism, (2) repoliticisations—recognising and articulating power relations, (3) for interdisciplinarity, (4) interventions—praxis and political engagement with research and (5) overcoming localism and spatial determinism.

Structure and Content

The book is divided into five parts, where each part discusses one of the above-mentioned themes. The chapters in each part are meant to allow scholars who have previously dealt with social acceptance of RET to critically reflect upon the research area that has often uncritically endorsed an approach of energy transition, embedded in neoliberal principles of green growth and energy substitution, but also to gather new insights from social studies of energy that have emerged at the fringes of the social acceptance research and beyond.

The need to *overcome individualisation and socio-cognitive reductionism* in research on the social acceptance of RET has been identified as relevant to further understand people's relations with RET for already some time, and is often coupled with the debate about the usefulness and consequences of using quantitative methods as opposed to qualitative ones (see Ellis et al., 2007, for a discussion). However, despite these calls, research on the social acceptance of RET, while within the neoliberal zeitgeist, has often tended to understand people's responses to RET as responses of individuals, conceived as either rationally calculating the costs and benefits of having RET deployed in the vicinity of where they live, or automatically and heuristically processing such decision-making processes inside their minds—or both. This is particularly present in research applying choice experiments and willingness-to-pay approaches (e.g. Ladenburg, 2009) to assess people's attitudes towards RET, which not only aim at an economisation of distance between (accumulated) individuals and RET, but also implicitly and uncritically tend to take NIMBY—Not in my backyard—rationales of the public for granted.

As pointed out by Kirkegaard and Nyborg (Chap. 6) in regard to a different theme, this relates with the fact that disciplines like social psychology and behavioural economics have significantly contributed to this area of research, while endorsing such individualist and positivist epistemological positions—even if not exclusively (see Batel et al., 2016; also Normann, 2020). In turn, those epistemological positions neglect how people's responses to RET are socially, materially, historically constructed, within networks of diverse human and non-human actors and associated power relations. The chapter by Kate Sherren, Ellen Chappell and John Parkins (Chap. 2) reflects on this theme and moves this discussion forward by proposing how quantitative methods, such as surveys and choice experiments, do not need to be necessarily designed within a positivist and individualist perspective, but can instead depart from these and gather contextually and socially relevant insights following more socio-constructivist underpinnings. Based on their research experience with energy transitions in Canada, the authors give several examples of tools and methods that can be used for that. Among these is Q-methodology, which had already been pointed out by Ellis and colleagues in 2007 as a way to methodologically overcome the positivism of traditional methods, and that is also used and discussed by Cuppen and Pesch (Chap. 9) in their analyses and practical insights on energy transitions in the Netherlands. In turn, Batel and Devine-Wright (Chap. 3) review the literature on the social acceptance of RET in relation to the role of the visual impacts of RET on people's responses, for highlighting how RET's visual impact has been increasingly estimated, measured, monetised or commodified, in short, by energy developers, policy-makers and researchers alike. The authors advocate that the visual-spatial impact of RET, instead of being taken as a consumer product, should rather be conceived as the dynamic and contingent node between people's life-place biographies, people-place relations and associated contextual fit of RET, and landscape traditions and associated ideologies. As such, the two chapters in this first part highlight that only by overcoming individualism and socio-cognitive reductionism in research on the social acceptance of RET we are able to uncover and address the power relations that shape social acceptance.

Power relations become the main focus of the second part of the book, *On repoliticisations—recognising and articulating power relations*, which aims to recover the political in research on the social acceptance of RET, often dismissed in this era of post-political green growth (Barry & Ellis, 2011; Swyngedouw, 2010; Uzzell & Räthzel, 2009). An often depoliticised and one-sided representation of RET obscures the social, political and economic processes underlying their development and diminishes issues of acceptance as purely localised matters as if they are not brought into being by an interplay of different power relations, contestations and political decisions at various scales (Batel, 2018; Kenis & Lievens, 2016). This necessitates a critical reflection of techno-managerialism as well as a discussion and examination of how energy infrastructures create, reflect or contest environmental (in)justices (Walker, 2009), energy colonialism (Batel & Devine-Wright, 2017), stigma (Rudolph & Kirkegaard, 2019) or neoliberal governmentality (Gailing & Leibenath, 2017). The chapter by Rudolph and Clausen (Chap. 4), which opens this part, departs from the famous U-curve assumption about social acceptance of RET to problematise the socio-spatial manifestations after RET projects have been installed and expose how a widely depoliticised post-construction phase within research on social acceptance of RET may hide issues related to a (lack of) citizenship, engagement and justice. In doing so, the authors argue that it is crucial to further consider and empirically analyse socio-spatial implications of RET after their installation and to provide thought-provoking impulses for reconsidering pre-emptive approaches to planning with ideas from radical planning. The following chapter by Alexander Dunlap (Chap. 5) takes the theme of this part to its most critical edge in his analysis of research on the social acceptance of RET, by proposing that we cannot talk about renewable energy as it is being deployed within current capitalist relations, but can only instead talk about fossil fuel+. As such, the author interestingly and provocatively discusses the entanglements of current modes of deploying RET and all the other inequalities, injustices and forms of oppression often characteristic of current socio-economic and political structures, and vividly narrates and illustrates them.

These chapters are followed by a part arguing *for interdisciplinarity*, which highlights why, albeit in different ways, in order to uncover and contest those entanglements and inequalities of current societal structures and associated human/non-human relations, it is crucial to adopt more interdisciplinary perspectives and develop more research inspired by different disciplines. The relevance of doing this in research on the social acceptance of RET is very well illustrated by Kirkegaard and Nyborg (Chap. 6), who argue for the relevance of bringing Actor Network Theory approaches to this field of research in order to open up, examine and understand unquestioned 'black boxes' of RET, such as related objects and materialities—the wind resource, stakeholders, sites—that are often taken for granted and unexplored, even if they are key to a better understanding of how RET, social acceptance and associated power relations are established and maintained. Also Claire Haggett (Chap. 7) discusses interdisciplinarity by highlighting the importance of research on the social acceptance of RET to better reflect about the terminology used in research, specifically, the meanings of attitude and behaviour, which have crystallised as key concepts in research on social acceptance through the years. What do attitude and behaviour mean in and for different disciplines, such as social psychology, economics and sociology? What implications might arise for research on the social acceptance of RET when adopting certain meanings instead of others? The author then makes the case, following up on previous calls on the importance of discourse and rhetoric in this area (Batel et al., 2013; Burningham, 2000), for the need of researchers to be more aware of the concepts they use, how they use them and with what consequences.

The next part, *On interventions—praxis and political engagement with the research*, makes explicit that all research is political while endorsing or contesting specific ways of representing people, energy infrastructures and other non-human entities and their relations (Batel et al., 2013; Shove, 2010). Nevertheless, most research so far has only engaged with what can/will be their practical and societal consequences in a very abstract and detached way. One of the main prerequisites of a critical approach to the study of people's responses to energy infrastructures is therefore to recognise and incorporate in praxis the political nature of all research questions asked and research designs conducted and scrutinise how the associated results are co-constructed, shared and appropriated by society at large. Nadaï and Labussière's chapter (Chap. 8) clearly illustrates how research is shaped by and shapes the encounters we have within fieldwork and with those affected by and connected with decisions about RET, as much as by our academic encounters, with books, colleagues and funding agencies. The authors nicely trace how their practical, conceptual and political engagements with energy transitions in France was co-constructed and changed throughout the years as part of the network created by those different encounters and how this experience prompted them to call for more academic engagement with collaborative experiments with communities and other human and non-human actors that compose these networks. Also Cuppen and Pesch (Chap. 9) interestingly reflect upon their long experience with participating in and thinking about the energy transition in the Netherlands and identify three main axes that should be further considered by research on the social acceptance of RET in order to make research more politically engaged and connected with the practical concerns and positionings of energy citizens: social conflict over RET as a multi-actor process, as a process of participation and as a dynamic process of interrelated events. As such, they again highlight, as we will further discuss in the Discussion part of this book, the relevance that relational approaches to the social acceptance of RET are gaining, as a consequence of both theoretical engagement in academic debates and empirical insights from fieldwork.

Finally, the last part illustrates several of the critical dimensions discussed so far by applying them in different socio-spatial and cultural contexts and associated energy transitions. Research on the social acceptance of energy infrastructures has often been criticised for its focus on the local, that is, on the analysis of local communities' responses to the siting of particular energy infrastructure projects (Ellis et al., 2007) and the neglect of how they are interrelated with activities at regional, national and global levels (Batel & Devine-Wright, 2017). This has led to argumentative struggles between universality of globalist views of climate change and reactionary particularism of local opposition to renewables, whereas both reproduce globalist or secluded privatised neoliberal understandings of energy landscapes (Mels, 2014). Although local particularities have been highlighted (e.g. Fast et al., 2015; Gee, 2010; Rudolph, 2014), the socio-spatial, cultural and structural conditions of the studied locales are still frequently overlooked, which necessitates a clearer consideration of the underlying energy transition processes in different contexts in order to overcome localism and spatial determinism. As such, this last part of the book contains chapters that specifically highlight that social acceptance of RET is not something constrained in local communities of nation-states in the Global North, but also happening in other areas of the world, where energy transitions take specific forms and reveal particular structural power relations while likewise causing impacts and dispossessions. The first chapter in this part by Gismondi and Hanson (Chap. 10) reveals the struggles that actors who advocate community energy projects face when clashing with centralised corporate-controlled energy politics in the Canadian province of Alberta that is characterised by an incumbent traditional fossil fuel economy. They show how the implementation of values related to renewable energy transition are hampered by incumbent actors and encounter a backlash from larger institutional and socio-historical contexts. The other two chapters in this part provide perspectives on social acceptance of RET from the Global South. The second chapter by Improta and Di Masso (Chap. 11) provides an analysis of discursive practices of local acceptance of a wind farm in Southern Brazil. They illustrate how local place-bonds are mobilised to legitimise the wind farm, while echoing historical colonial understandings of progress and modernisation. Based on their findings, the authors argue for the consideration of situated knowledge, local empowerment and locally driven development instead of technology and policy transfer from the Global North. The last chapter by van der Horst and colleagues (Chap. 12) assesses the efficacy of the energy justice framework to make sense of social acceptance of renewable energy for electrification purposes in Mexico and Malawi. In doing so, the authors bring together the fields of energy justice and social acceptance that have emerged in the Global North and discuss their relevance in the context of the Global South, illustrate the conditionalities of nuanced injustices and argue for a contextually rooted understanding of social acceptance. As such, they again highlight the significance of the social, political and historical context in shaping the development of renewables and the issue of acceptance.

While addressing and discussing one of the five themes that we have proposed as key to a critical approach to the social acceptance of RET, each chapter also contributes to an illustration of what this critical approach may mean and to show that some of its dimensions create tensions and are open to diverse uptakes, often reflecting different disciplinary backgrounds and different political and normative positionings and considerations. As such, each chapter also made us further reflect on this critical approach and the new directions it can and should take in shaping research on the social acceptance of RET in particular and on energy transitions more generally. Thus, revisiting each chapter's contribution to a critical approach, the reflection and discussion of the tensions and different interpretations they reveal, as well as how they contribute to shaping new directions of inquiry in adopting a critical approach to research on the social acceptance of RET, will be part of the Discussion presented at the end of the book.

References

Aitken, M. (2010). Wind power and community benefits: Challenges and opportunities. *Energy Policy, 38*(10), 6066–6075.

Aitken, M., Haggett, C., & Rudolph, D. (2016). Practices and rationales of community engagement with wind farms: Awareness raising, consultation, empowerment. *Planning Theory & Practice, 17*(4), 557–576.

Barry, J., & Ellis, G. (2011). Beyond consensus? Agonism, republicanism and a low carbon future. In *Renewable energy and the public* (pp. 61–74). Routledge.

Batel, S. (2018). A critical discussion of research on the social acceptance of renewable energy generation and associated infrastructures and an agenda for the future. *Journal of Environmental Policy & Planning, 20*(3), 356–369.

Batel, S. (2020). Research on the social acceptance of renewable energy technologies: Past, present and future. *Energy Research & Social Science, 68*, 101544.

Batel, S., & Adams, M. (2016). Ecological crisis, sustainability and social worlds: Developing a critical agenda. *Papers on Social Representations, 25*(1), 1.1–1.27.

Batel, S., Castro, P., Devine-Wright, P., & Howarth, C. (2016). Developing a critical agenda to understand pro-environmental actions: Contributions from Social Representations and Social Practices Theories. *Wiley Interdisciplinary Reviews: Climate Change, 7*(5), 727–745.

Batel, S., & Devine-Wright, P. (2017). Energy colonialism and the role of the global in local responses to new energy infrastructures in the UK: A critical and exploratory empirical analysis. *Antipode, 49*(1), 3–22.

Batel, S., Devine-Wright, P., & Tangeland, T. (2013). Social acceptance of low carbon energy and associated infrastructures: A critical discussion. *Energy Policy, 58*, 1–5.

Baxter, J., Walker, C., Ellis, G., Devine-Wright, P., Adams, M., & Fullerton, R. S. (2020). Scale, history and justice in community wind energy: An empirical review. *Energy Research & Social Science, 68*, 101532.

Becker, S., & Naumann, M. (2017). Energy democracy: Mapping the debate on energy alternatives. *Geography Compass, 11*(8), e12321.

Bosch, S., & Schmidt, M. (2019). Is the post-fossil era necessarily post-capitalistic? – The robustness and capabilities of green capitalism. *Ecological Economics, 161*, 270–279.

Brand, U., & Wissen, M. (2018). *The limits to capitalist nature: Theorizing and overcoming the imperial mode of living*. Rowman & Littlefield.

Bridge, G., Barr, S., Bouzarovski, S., Bradshaw, M., Brown, E., Bulkeley, H., & Walker, G. (2018). *Energy and society: A critical perspective*. Routledge.

Burningham, K. (2000). Using the language of NIMBY: A topic for research, not an activity for researchers. *Local Environment, 5*(1), 55–67.

Clausen, L. T., & Rudolph, D. (2020). Renewable energy for sustainable rural development: Synergies and mismatches. *Energy Policy, 138*, 111289.

Cowell, R., Bristow, G., & Munday, M. (2011). Acceptance, acceptability and environmental justice: The role of community benefits in wind energy development. *Journal of Environmental Planning and Management, 54*(4), 539–557.

Creamer, E., Aiken, G. T., van Veelen, B., Walker, G., & Devine-Wright, P. (2019). Community renewable energy: What does it do? Walker and Devine-Wright (2008) ten years on. *Energy Research & Social Science, 57*, 101223.

Creamer, E., Eadson, W., van Veelen, B., Pinker, A., Tingey, M., Braunholtz-Speight, T., Markantoni, M., Foden, M., & Lacey-Barnacle, M. (2018). Community energy: Entanglements of community, state, and private sector. *Geography Compass, 12*(7), e12378.

Daggett, C. (2020). Energy and domination: Contesting the fossil myth of fuel expansion. *Environmental Politics, 30*(4), 644–662.

Devine-Wright, P. (2011). *Renewable energy and the pubic: From NIMBY to participation*. Routledge.

Dunlap, A. (2018). The 'solution' is now the 'problem': Wind energy, colonisation and the 'genocide-ecocide nexus' in the Isthmus of Tehuantepec, Oaxaca. *The International Journal of Human Rights, 22*(4), 550–573.

Dunlap, A. (2020). Wind, coal, and copper: the politics of land grabbing, counterinsurgency, and the social engineering of extraction. *Globalizations, 17*(4), 661–682.

Eadson, W., & Foden, M. (2019). State, community and the negotiated construction of energy markets: Community energy policy in England. *Geoforum, 100*, 21–31.

Ellis, G., Barry, J., & Robinson, C. (2007). Many ways to say 'no', different ways to say 'yes': Applying Q-methodology to understand public acceptance of wind farm proposals. *Journal of Environmental Planning and Management, 50*(4), 517–551.

Fast, S., Mabee, W., & Blair, J. (2015). The changing cultural and economic values of wind energy landscapes. *The Canadian Geographer/Le Géographe canadien, 59*(2), 181–193.

Fraune, C., Knodt, M., Gölz, S., & Langer, K. (2019). *Akzeptanz und politische Partizipation in der Energietransformation. Gesellschaftliche Herausforderungen jenseits von Technik und Ressourcenausstattung* [Acceptance and political participation in energy transformation. Societal challenges beyond technology and resources]. Springer.

Gailing, L., & Leibenath, M. (2017). Political landscapes between manifestations and democracy, identities and power. *Landscape Research, 42*(4), 337–348.

Gee, K. (2010). Offshore wind power development as affected by seascape values on the German North Sea coast. *Land Use Policy, 27*(2), 185–194.

Giotitsas, C., Nardelli, P. H., Kostakis, V., & Narayanan, A. (2020). From private to public governance: The case for reconfiguring energy systems as a commons. *Energy Research & Social Science, 70*, 101737.

Gorroño-Albizu, L., Sperling, K., & Djørup, S. (2019). The past, present and uncertain future of community energy in Denmark: Critically reviewing and conceptualising citizen ownership. *Energy Research & Social Science, 57*, 101231.

Gross, C. (2007). Community perspectives of wind energy in Australia: The application of a justice and community fairness framework to increase social acceptance. *Energy Policy, 35*(5), 2727–2736.

Haraway, D. (2015). Anthropocene, capitalocene, plantationocene, chthulucene: Making kin. *Environmental Humanities, 6*(1), 159–165.

Harris, J. (2011). Going green to stay in the black: Transnational capitalism and renewable energy. *Perspectives on Global Development and Technology, 10*(1), 41–59.

Holstenkamp, L., & Radtke, J. (2018). *Handbuch Energiewende und Partizipation* [Handbook energy transition and participation]. Springer.

Howe, C., & Boyer, D. (2016). Aeolian extractivism and community wind in Southern Mexico. *Public Culture, 28*(2 (79)), 215–235.

Jørgensen, M. L. (2020). Low-carbon but corrupt? Bribery, inappropriateness and unfairness concerns in Danish energy policy. *Energy Research & Social Science, 70*, 101663.

Kenis, A., & Lievens, M. (2015). *The limits of the Green Economy: From re-inventing capitalism to re-politicising the present*. Routledge.

Kenis, A., & Lievens, M. (2016). Greening the economy or economizing the green project? When environmental concerns are turned into a means to save the market. *Review of Radical Political Economics, 48*(2), 217–234.

Kirkegaard, J., Cronin, T., Nyborg, S., & Karnøe, P. (2020). Paradigm shift in Danish wind power: The (un)sustainable transformation of a sector. *Journal of Environmental Policy & Planning*, 1–17.

Lacey-Barnacle, M. (2020). Proximities of energy justice: Contesting community energy and austerity in England. *Energy Research & Social Science, 69*, 101713.

Ladenburg, J. (2009). Stated public preferences for on-land and offshore wind power generation – A review. *Wind Energy: An International Journal for Progress and Applications in Wind Power Conversion Technology, 12*(2), 171–181.

Lennon, B., Dunphy, N., Gaffney, C., Revez, A., Mullally, G., & O'Connor, P. (2020). Citizen or consumer? Reconsidering energy citizenship. *Journal of Environmental Policy & Planning, 22*(2), 184–197.

MacLeod, M. A., & Emejulu, A. (2014). Neoliberalism with a community face? A critical analysis of asset-based community development in Scotland. *Journal of Community Practice, 22*(4), 430–450.

McCarthy, J. (2015). A socioecological fix to capitalist crisis and climate change? The possibilities and limits of renewable energy. *Environment and Planning A, 47*(12), 2485–2502.

Mels, T. (2014). Globalism, particularism, and the greening of neoliberal energy landscapes. In *Green utopianism. Perspectives, politics and micro-practices* (pp. 165–179). Routledge.

Natarajan, L., Rydin, Y., Lock, S. J., & Lee, M. (2018). Navigating the participatory processes of renewable energy infrastructure regulation: A 'local participant perspective' on the NSIPs regime in England and Wales. *Energy Policy, 114*, 201–210.

Normann, S. (2020). Green colonialism in the Nordic context: Exploring Southern Saami representations of wind energy development. *Journal of Community Psychology, 49*(1), 77–94.

Pepermans, Y., & Maeseele, P. (2014). Democratic debate and mediated discourses on climate change: From consensus to de/politicization. *Environmental Communication, 8*(2), 216–232.

Raman, S. (2013). Fossilizing renewable energies. *Science as Culture, 22*(2), 172–180.

Rudolph, D. (2014). The resurgent conflict between offshore wind farms and tourism: Underlying storylines. *Scottish Geographical Journal, 130*(3), 168–187.

Rudolph, D., Haggett, C., & Aitken, M. (2018). Community benefits from offshore renewables: The relationship between different understandings of impact, community, and benefit. *Environment and Planning C: Politics and Space, 36*(1), 92–117.

Rudolph, D., & Kirkegaard, J. K. (2019). Making space for wind farms: Practices of territorial stigmatisation in rural Denmark. *Antipode, 51*(2), 642–663.

Shove, E. (2010). Beyond the ABC: Climate change policy and theories of social change. *Environment and Planning A, 42*(6), 1273–1285.

Slee, B. (2015). Is there a case for community-based equity participation in Scottish on-shore wind energy production? Gaps in evidence and research needs. *Renewable and Sustainable Energy Reviews, 41*, 540–549.

Stober, D., Suškevičs, M., Eiter, S., Müller, S., Martinát, S., & Buchecker, M. (2021). What is the quality of participatory renewable energy planning in Europe? A comparative analysis of innovative practices in 25 projects. *Energy Research & Social Science, 71*, 101804.

Strachan, P. A., Cowell, R., Ellis, G., Sherry-Brennan, F., & Toke, D. (2015). Promoting community renewable energy in a corporate energy world. *Sustainable Development, 23*(2), 96–109.

Swyngedouw, E. (2010). Apocalypse forever? Post-political populism and the spectre of climate change. *Theory, Culture & Society, 27*(2–3), 213–232.

Swyngedouw, E. (2011). Interrogating post-democratization: Reclaiming egalitarian political spaces. *Political Geography, 30*(7), 370–380.

Upham, P., Bögel, P., & Johansen, K. (2019). *Energy transitions and social psychology: A sociotechnical perspective*. Routledge.

Uzzell, D., & Räthzel, N. (2009). Transforming environmental psychology. *Journal of Environmental Psychology, 29*(3), 340–350.

van Veelen, B. (2020). Caught in the middle? Creating and contesting intermediary spaces in low-carbon transitions. *Environment and Planning C: Politics and Space, 38*(1), 116–133.

Walker, G. (2009). Beyond distribution and proximity: Exploring the multiple spatialities of environmental justice. *Antipode, 41*(4), 614–636.

Warren, C., Cowell, R., Ellis, G., Strachan, P. A., & Szarka, J. (2012). Wind power: Towards a sustainable energy future? In *Learning from wind power* (pp. 1–14). Palgrave Macmillan.

Wüstenhagen, R., Wolsink, M., & Bürer, M. J. (2007). Social acceptance of renewable energy innovation: An introduction to the concept. *Energy policy, 35*(5), 2683–2691.

PART II

Overcoming Individualism and Socio-Cognitive Reductionism

CHAPTER 2

Strategies for Integrating Quantitative Methods into Critical Social Acceptance Research

Kate Sherren, Ellen Chappell, and John Parkins

INTRODUCTION

Conflicts over energy technologies and infrastructure development, and the influence of those conflicts on decision-making, demonstrate that renewable energy transitions are vulnerable to public opinion. We thus need a good understanding of public opinion and its influences to advance such transitions. Various terms are used in the literature to describe such opinions, including acceptance, opposition, support, with calls for more nuanced conceptualizations (Batel et al., 2013). Quantitative methods such as household surveys are conventionally used to gauge public

K. Sherren (✉) • E. Chappell
Dalhousie University, Halifax, NS, Canada
e-mail: kate.sherren@dal.ca

J. Parkins
University of Alberta, Edmonton, AB, Canada
e-mail: jparkins@ualberta.ca

S. Batel, D. Rudolph (eds.), *A critical approach to the social acceptance of renewable energy infrastructures*,
https://doi.org/10.1007/978-3-030-73699-6_2

opinion and understand its patterns and drivers at a population scale. Well-designed, implemented and analysed surveys can enable researchers to identify significant patterns and draw inferences. But without an experimental design such methods can struggle to identify causality. Indeed, what Firestone (1987) calls the "hydraulic forcing" of individual-scale variables assumed to act upon predicted outcomes is often subject to the hypothesized causality at play. Interview methods can do this better, with smaller numbers of people, by simply asking questions about "why" such opinions are held by the research participant. Commensurate with the shift from surveys to more qualitative critical research around the acceptability of renewable energy technology (RET), the deployment of quantitative surveys has been subject to criticisms from those engaged in critical energy research (Batel & Devine-Wright, 2015; Ellis et al., 2007). This critique extends to other fields affected by the replicability crisis and research norms that enable and even encourage question re-use and data-driven analysis over theory-driven work (Camerer et al., 2018; Wicherts et al., 2016). Nevertheless, we believe survey work on energy acceptability can indeed be critical and can overcome some of the criticisms noted here.

What do we mean by critical? Borrowing from Morrow and Brown (1994) in the social sciences, the term critical itself is concerned with "unveiling ideological mystifications in social relations", establishing a critical stance towards methodology as a vehicle for "understanding the nature of reality, knowledge, and explanation" and the "self-reflexivity of the investigator and the linguistic basis of representation" (p. 7). These issues are at the heart of what is understood to be a critical social science. For the purposes of this chapter, we focus on the methodological questions involving the nature of reality and explanation. On one level, we sense that researchers within the qualitative and interpretivist traditions of social science find quantitative approaches to be uncritical in part because much of what passes as quantitative analysis is little more than descriptive statistics. Reporting frequencies, distributions and average levels of support or opposition to new wind farms is hardly social science and is at best straightforward opinion polling. Moreover, many interpretive scholars find unsatisfying the reduction of complex social settings to sets of numbers. Yet, we seek to illustrate through this chapter a thread of insight into the potential for critical inquiry in quantitative social science. We define our work here as critical social science because it attempts to address complex social problems in complex ways, moves beyond conventional question sets and considers themes that capture context and nuanced understandings of reality and explanation. Within survey research we are typically limited to the single respondent, but this chapter outlines several ways to move beyond methodological individualism and conventional questions to engage with the many ways that people experience RET, integrate it into their landscapes, livelihoods and relationships, and how this plays out in a complex multi-dimensional context. We make our case by drawing on innovations we employed in three recent energy-related surveys in Canada, at three different scales of nation, province and region, and by building on related literature.

Methods

While this is not an empirical chapter, a brief description of the three surveys on which we draw most is warranted. These surveys have been undertaken as part of the Energy Transitions in Canada project led by John Parkins since 2012 and supported by the Social Sciences and Humanities Research Council of Canada. Each was developed for different purposes and populations and implemented in different ways. They are examples of national, provincial and regional surveys, respectively.

The first survey was a 2014 proportional survey of Canadians (n = 3000), implemented by polling firm Corporate Research Associates using online panels, with an n = 500 oversample of the small province of New Brunswick used to better understand attitudes towards a failing hydroelectric dam there (Sherren et al., 2017). This national survey explored support for a dozen different energy sources along with related phenomena (e.g. solar adoption (Parkins et al., 2018)), self-reported exposure to energy infrastructure (Sherren et al., 2019), social and institutional trust (Parkins et al., 2017) and other issues such as energy literacy and environmental values. This survey is referred to below as the national survey.

The second survey was conducted in early 2019 with a focus on large-scale landholders in the province of Alberta. Survey respondents (n = 401) are those who would be most well-placed to host RET as the province was ramping up renewable electricity production. We secured an online panel for this survey through the international agricultural marketing firm, Kynetec. Research focused on the conditions associated with support and opposition to wind power development in the province, and our methods involved a factorial experiment that is discussed in more detail below. This survey is referred to below as the provincial survey.

The third survey was a 2018 regional survey of residents of the Chignecto isthmus that connects the small provinces of New Brunswick and Nova Scotia in Canada's east (n = 335). This multi-reminder mailout survey achieved a response rate of 40% and was implemented to explore perceptions of landscape change (Chappell et al., 2020) and wind energy (Chappell et al., in press) in the rural area, taking advantage of the recent construction of a 15-turbine wind farm around the same time and near the same place that a large microwave radio tower array was decommissioned and removed. The survey used a photo-based experimental treatment that is discussed in more detail below. This survey is referred to below as the regional survey.

Beyond Methodological Individualism

With a strong focus on beliefs, values and attitudes, much of the survey research on social acceptability remains focused squarely on the individual respondent and their belief system (Batel, 2018; Schlüter et al., 2017). Indeed, because surveys are often done in the respondent's own home, whether they were recruited by a polling firm or they randomly received a survey by mail or online, researchers cannot be sure who is actually participating nor what level of care they bring to that task. In our other work on agriculture, for instance, we have had reason to question whether the person who makes the most farm decisions is actually filling out surveys about farm management, as requested, or the person who does their paperwork (Goodale et al., 2015). Notwithstanding this fact, there are ways within surveys to explore the social, cultural and environmental context in which that individual resides and how that might shape their thinking and behaviour.

Engaging with Place and Materiality

An individual's opinions about RET depends in part upon their physical setting, whether it currently includes energy infrastructure or not, and the nature of their experiences with that setting or place. Such a rich story can only be partially explored in surveys, rather than *in situ* or photo elicitation interviews for instance, but surveys can provide important insight. Some research captures this context through research design, by sampling areas around wind farms (Walker et al., 2014) or using postal codes to identify respondents in proximity to features of interest (Clarke et al., 2016). Yet people vary in their tendency to notice such infrastructure or assign it to its actual purpose (e.g. energy production rather than manufacturing, for instance). Moreover, it is likely not only the experience at one's mailing address that matters when it comes to RET support. Instead of—or in addition to—such implied exposure, it is important to ask about exposure, perhaps including photographs or other stimuli for clarity. It may also be useful not to limit questions about exposure to the purely visual; senses like smell and hearing are also relevant influences on RET support (Fergen & Jacquet, 2016; Firestone et al., 2015). Such questionnaire content will provide a richer picture than that advanced by use of simplistic NIMBY models and help RET proponents to design more appropriate modes of engagement with potential host communities (Devine-Wright, 2011).

We have explored this materiality in our surveys. In the regional survey we provided a wind turbine photo and asked if respondents had seen or heard them before, and—as expected—most had. Some respondents indicated they could see but not hear the turbines. But we also asked a range of questions to establish how often the respondent saw wind turbines and from where (e.g. home, work or school), to understand the extent and pattern of that exposure (Chappell et al., 2020, in press). In the national survey we also addressed materiality through multiple senses, asking—about a dozen energy infrastructure types—"*Which of the following do you regularly, see, hear or smell? [Select all that apply] Hydroelectric dam, wind turbines, nuclear plant, coal-fired plant, etc.*" This list of sensory experiences was important given the impact of infrasound or air quality impacts on the perceived acceptability of many energy sources. Although many respondents were in physical proximity to such infrastructure, a third of respondents indicated they did not experience any energy infrastructure (including transmission lines). These respondents either filtered out that infrastructure or simply did not recognize it. It is important to be able to separately analyse actual and perceived exposure to RET, such as by combining data on household proximity to RET and survey content discussed above. Believing one is exposed when one is not may drive opinion as much as actually being exposed and believing one is not exposed when one is exposed tells us much about landscape norms and expectations (Sherren et al., 2019).

Looking Relationally

The back page of any survey is usually saved for demographic variables about the respondent that can be used as controls or predictors of energy support and/or acceptance. These variables often include gender, age, income, education, employment status and political beliefs. It is rarer to ask respondents about their relationships with, dependencies upon or levels of trust in others. Stern and colleagues (Stern & Baird, 2015; Stern & Coleman, 2015) define trust as the willingness to be vulnerable to an entity while in a state of uncertainty, which is critical to shaping many energy resource and siting decisions. Trust, lack of trust or distrust are distinct states and theorized as emerging from characteristics of the trustor (the person trusting), the trustee (what is being trusted) and the context (Stern & Coleman, 2015); all of these can be queried in surveys. There is a relational turn underway in human-environment research that locates values or subjective states such as resilience as emergent between people or between people and environments, a product of their interactions, rather than residing with any one of them *per se* (Chan et al., 2016; Darnhofer et al., 2016; Grubert, 2018). Trust emerges in large measure through histories of interaction between people and groups and has been identified by many researchers as a central element of RET acceptance (Hall et al., 2013; Rand & Hoen, 2017). Trust and norms also emerge in part from relationships and comparisons with others, whether those are direct interactions with those socially or geographically proximate, or indirect comparisons with those more distant, such as opinion leaders or celebrities. Relationships and dependencies can thus be used as vectors for opinions about energy, as people seek cognitive shortcuts to avoid overload in busy lives (Iftekhar & Pannell, 2015). This relational turn is an invitation in our survey methods to look past the respondent and include social variables that indicate how the respondent is socially located, how their livelihood is sustained and the levels of trust that exist between that individual and a range of relevant institutions and organizations.

In the national survey we asked whether respondents or their household/family members were employed in a range of different aspects of the energy sector, including each specific industry (e.g. oil or wind) as well as energy-related government departments and regulatory bodies ("*Do you personally /Does any household or family member work for any of the following energy-related industries?*"). The employment data demonstrated how much support for fossil-fuel-based energy sources was linked to family jobs in specific energy industries, a dramatic contrast with renewable energy where employment was unrelated (Sherren et al., 2019). This impact was most dramatic for coal, where having a family member employed in the sector almost quadrupled the likelihood that the respondent would support coal.

In the same national survey, we had several questions that explored social and institutional trust ("*How much do you trust or distrust the following as sources of information on energy issues?*"). These trust questions allowed us to explore the benefits but also the risks of trust, particularly the degree to which too much trust can lead to apathy in public engagement on energy issues (Parkins et al., 2017). When exploring gender differences in support for coal, using the national survey dataset, we found that trust variables were extremely strong predictors of women's opinions on coal, specifically whether they trusted industry insiders or outsiders such as academics and non-government organizations (Parkins et al., 2020). These questions on dependency and trust represent an important expansion from the views and preferences of the respondent to the social context and relationships in which the respondent is embedded.

Exploring Situated Norms

Using household survey techniques alone, it is challenging to move beyond the individual to group-level behaviour or collective thinking because we are restricted to the aggregation of individual responses. Nonetheless, for a long time, researchers have recognized the role of collective norms on individual beliefs and behaviours (Farrow et al., 2017). The literature makes it clear that much of our individual behaviour is in direct response to social norms. Paluck (2009) defines social norms as socially shared beliefs prescribing or proscribing social behaviours, whereas Ehrlich and Levin (2005) described them not only as rules, but also as typical behavioural patterns within social groups. There is, nonetheless, general consensus that these norms, either as beliefs, rules or behaviours, are communicated to individuals through shared interaction by the use of positive or negative sanctions and through messaging transmitted via media and other forms of communication. Types of norms are also evident. For example, researchers distinguish between *injunctive* norms and *descriptive* norms; injunctive norms representing an individual's interpretation of behaviours that *ought* to be followed and descriptive norms representing an interpretation of behaviours that *are* being followed by the majority. The significance of this distinction is that the perceived existence of sanctions for nonconformity will have an enhanced effect on an individual's perception or behaviour if there is also a perception that the majority of others conform to that norm (Biel & Thøgersen, 2007).

To date, energy social science has paid limited attention to the role of norms in the formation of support or opposition to energy development. This may be, in part, because people tend to underestimate and/or under-report the degree to which they are influenced by others (Stern, 2018). Yet the literature on this topic is compelling and we sense a need to utilize questions about norms as a way to move beyond the potential atomization of individual experience and preference and to delve more deeply into the collective experiences and resulting expectations of people within their geographic and epistemic communities, what are called situated norms. Within the provincial survey, we included a set of variables related to such norms (Holowach et al., 2019). In response to each statement listed as follows, respondents were invited to agree or disagree on a scale from 1 to 5:

- *People here are indifferent about supporting local enterprises.*
- *For the most part, my local community would be excited about a new wind farm.*
- *I would be considered rude if I didn't talk to my neighbours before making decisions about my land that could affect them.*
- *Farmers in this county greatly disapprove of people who take more than their fair share.*
- *Poor stewardship of one's land is greatly frowned upon here.*
- *In this community, it doesn't matter as much about how a decision is made, rather only that the outcome is fair.*

These statements reflect a subtle but important shift away from individual-level perspectives on identity (e.g. my land is a big part of my identity) or attitudes (e.g. I am concerned about climate change) to a set of statements that signal something about the collective, particularly as they relate to a set of descriptive norms about what the community values and how neighbours might respond to specific events, such as RET development. Moving forward from these initial efforts, we propose deeper engagement with the assessment of norms and more formal approaches to measuring norms that are now becoming more prominent in the literature (Bicchieri, 2017). Though most quantitative researchers conceive of such variables as external predictors of a given attitude or behaviour, it is important to be circumspect about assuming any particular direction of causality when discussing such relationships: norms, attitudes and behaviours are deeply entangled and recursively reproduced (Giddens, 1979).

Beyond Likert Questions

Survey questions take many forms and there are strong disciplinary traditions in terms of how questionnaires are constructed. Within psychology we observe a focus on established and repeated use of scales or new scale development for conditions such as depression or place attachment. Within economics, researchers focus on trade-offs and tools such as choice experiments to identify preferences for certain types of attributes (e.g. environmental outcomes) where prices or costs are not deduced from market transactions. Recognizing there is tremendous variety in survey types, we cannot overstate the importance of Likert scales across all disciplines. Responses to statements like "climate change is a grave and urgent threat", with response options that range from strongly agree to strongly disagree, are a cornerstone of psychometric research.

As individual items or as an aggregation of multiple items within a scale, this approach to quantitative research is not without its merits. Yet the overuse of such approaches to research limits our capacity to gain deeper insights into the social world and the subjective states of people and groups in question. For example, in studies of public attitudes towards new energy technologies, Likert scales give a sense of the relative preferences that people hold and a sense of the factors that are correlated with these preferences. But a simple analysis of frequencies or correlates can be misleading, in part because cross-sectional data offer limited opportunities to detect causality. More importantly, Likert scales offer no opportunity to assess trade-offs. When respondents are free to support all energy technologies or reject all renewable technologies, then the respondent is free to express a version of reality that is entirely unrealistic. A more realistic survey approach would compel the respondent to consider real-world trade-offs. For example, if you claim to support wind turbines, are you also willing to live with turbines in your everyday views or (depending on local electricity markets) are you willing to pay more for power? This approach to survey research offers a more realistic and critical perspective on RET acceptance, because it offers a more realistic picture of the social context and the complexity of social acceptability.

Forcing Trade-offs

There are two ways in which our research attempts to bring a consideration of trade-offs into the analysis. First, we utilize Q methodology as a tool for discovering political discourses associated with energy production. Political discourses are the embodied sets of "capabilities which enable the assemblage of words, phrases and sentences into meaningful 'text' intelligible to readers or listeners" (Dryzek & Berejikian, 1993, p. 51). Q methodology is frequently employed to provide a nuanced understanding of subjective value positions and interests around policy and planning choices. Specific procedures include a careful selection of 30–50 statements on a specific topic (e.g. energy production in Canada), representing the "universe of ideas" on the topic. Respondents then participate in a process of selecting statements they agree with more in relation to statements they agree with less, resulting in a distribution of preferences that are unique to each participant. Forcing trade-offs is an explicit component of this procedure. Finally, these results are subjected to principal component analysis resulting in a statistically derived set of underlying factors that explain variation within the dataset.

While Q methodology can be administered online and thus with larger groups (Sherren, Loik, & Debner, 2016), often the value of this method comes in working with smaller groups of people and identifying the ways in which otherwise very subjective positions on topics of concern results in systematic and statistically robust understandings of inherently subjective ideas (Mann et al., 2019). In our study of discourses on energy carried out in person in three provinces, we identified five discourses ranging from climate concerned to more nuanced understandings of resilience and local energy systems (Parkins et al., 2015). We used the factor-defining statements from that study in our national survey and found strong connections to the same discourses nationally (Parkins & Sherren, 2021). These insights into frames of thinking and reasoning can facilitate mutual understanding, recognition of shared values and opportunities for constructive dialogue.

Using Vignettes and Scenarios

A second strategy for avoiding simplistic Likert response variables involves the development of vignettes. Vignettes are designed to characterize and assess complex social settings and can be embedded within otherwise conventional cross-sectional survey designs. Defined as factorial survey experiments or vignette experiments (Auspurg & Hinz, 2014), the idea of social acceptability is considered to be a function of how an individual perceives a RET as well as the broader social context in which the technology is embedded. Who gets to decide how the project is designed? Who owns and controls the project? How is compensation determined for people living near the turbine? These questions are deeply connected to issues of social acceptability and can be examined more holistically through vignette experiments. Modelled on similar work in Poland and Germany (Liebe et al., 2017), our provincial study provided large-scale landowners, those who are most likely to host turbines on their land, with a set of scenarios that represent trade-offs in how a potential wind farm might be owned, designed, implemented and managed. In these scenarios we identified a series of factors that vary between scenarios, resulting in a total of 144 unique scenarios that allow an empirically optimal combination of individual settings for each factor to be tested. For each survey participant, six of these scenarios were randomly presented within the online questionnaire. As an example, one of these scenarios reads as follows:

> There is an opportunity for a <u>local cooperative</u> to develop a wind farm <u>on your neighbour's property</u> (5 km away). For projects like this, other residents living nearby will all <u>receive some compensation</u> based on their proximity to the turbines. All county residents will be <u>invited to express concern</u> about the project. Meanwhile, details about the lease payments and compensation amounts <u>will be available</u> to anyone directly affected by the project.

The underlined portions of the vignette above (not underlined in the survey) are the factors that vary between scenarios, offering a more complex approach to understanding the social context of support and opposition to wind power development. With preliminary analysis completed (Holowach et al., 2019), results show that landowners in Alberta prefer proximity-based compensation, in contrast to the standard approach which compensates the landowner through lease payments but offers nothing to neighbouring landowners. We also found a preference for local cooperative ownership of wind farms, which represents a governance model in western farming communities that is common for the distribution of electricity but remains rare for its production.

Enquiring About Emotions

Based on the analysis described above, much work is focused on distributional and procedural justice, but much less work is focused on how landscape changes and new infrastructure make people feel. The role of emotions in social sciences has gained new interest, particularly within sociology. Bericat (2016) suggests that if sociology is interested in explaining the fundamentals of social behaviour, then sociological analyses must continue to expand on and be inclusive of the role of emotions in diverse areas of inquiry. Furthermore, she identifies two distinct types of emotions,

> *Primary* emotions are considered to be universal, physiological, of evolutionary relevance and biologically and neurologically innate, while *secondary* emotions, which can be a result of a combination of primary emotions, are socially and culturally conditioned. (p. 492)

Emotions have thus far been avoided by many quantitative researchers in the energy space (exceptions include Maehr et al., 2015; Zaunbrecher et al., 2018), perhaps because of suspicions that they are ephemeral. All cross-sectional social science methods (by contrast with longitudinal) are vulnerable to ephemerality, and many variables can change day by day, such as how fair a siting process feels or even the level of support a respondent feels for a given technology. Ephemerality of such feelings is a human condition and does not negate their influence or importance.

Secondary emotions are at the heart of our recent work on historical landscape features, landscape change and the possibilities for repurposing landscape for RET. In the regional survey, we presented half of the respondents with photographs of past and current landscape changes (Fig. 2.1), where built utilitarian infrastructure had been or was being removed.

The experimental treatment was to see if being reminded of the extent of past change made people feel more open to future change such as increased wind turbine installation (it didn't), but it also allowed us to explore how legacy infrastructure inspired expressions of fit and sadness (Chappell et al., 2020). Males and conservatives were most likely to (a) feel that such infrastructure fit and (b) be sad at its loss, both perhaps related to the same underlying phenomena as petro-masculinities (Daggett, 2018). We also learned that those who tend to become attached to landscape features in this way are also likely to become attached to wind turbines and be open to more. Our exploration of emotion was a preliminary

4. Perceptions of past landscape changes: Please read the following information about previous landscape uses and features in the Chignecto area and then respond to the questions by ticking your response.

a. Dykes: Dykes were constructed by the Acadians in the 1600s and 1700s to drain the salt marsh and to create agricultural land. Today, some dykes are being moved or breached to return the salt marsh to its original habitat and to help protect against rising sea levels.

	Yes	No
Have you noticed the dykes in the Chignecto area? *(If no, skip to Question 4b.)*	○	○

	Disagree	Neutral	Agree
The dykes fit well in the Chignecto landscape.	○	○	○
I am sad at the loss of the dykelands.	○	○	○

b. Foundries: Foundries were established in the Chignecto area for metal processing in the mid-1800s. While some buildings remain, none of the tall smoke stacks remain.

	Yes	No
Did you live in the Chignecto area prior to the loss of the foundries? *(If not, skip to Question 4c.)*	○	○

	Disagree	Neutral	Agree
The foundries fit well in the Chignecto landscape.	○	○	○
I am sad to see the loss of the foundry stacks.	○	○	○

c. Hay barns: Hay barns were constructed through the 1800s and 1900s to store hay harvested from the marsh to feed horses in the area and for export. At one time there were more than 400 hay barns, but the few remaining today are collapsing.

	Yes	No
Have you noticed hay barns in the Chignecto area? *(If no, skip to Question 4d.)*	○	○

	Disagree	Neutral	Agree
The hay barns fit well in the Chignecto landscape.	○	○	○
I am sad to see the loss of the hay barns.	○	○	○

d. RadioTowers: The Radio Canada International (RCI) towers were built during WWII to transmit radio to Canada and overseas, and remained until the Internet largely replaced radio. They were dismantled in 2014.

	Yes	No
Did you live in the Chignecto area prior to the removal of the RCI towers? *(If not, skip to Question 5.)*	○	○

	Disagree	Neutral	Agree
The RCI towers fit well in the Chignecto landscape.	○	○	○
I am sad to see the loss of the RCI towers.	○	○	○

Fig. 2.1 Photo representations of landscape change in the Chignecto region; half of the respondents in our regional survey received this page that invited them to look at photographs of past landscape changes in the region and state whether they lived in the area at the time the features existed and then asked for an expression of fit, and whether the loss of them made the respondents sad

foray into this space, but such work could fill in some of the gaps that result from a focus on more rational assessments of, for instance, landscape impacts or procedural justice (Gross, 2007; Walker & Baxter, 2017).

Discussion and Conclusion

We set out in this chapter to illustrate some of the ways that we seek to carry out critical social science in our use of survey methods, by moving beyond methodological individualism and standard Likert question sets. This approach includes engagement with the sensory experience of energy, its relational dimensions and the situated norms that influence acceptance, as well as innovative question sets that force trade-offs, present narratives and explore emotions. We are deeply curious about the social reality at play and work creatively to explore them in surveys, if occasionally imperfectly or by use of proxies. This kind of reflexive experimentation is important for the field and for gaining deeper insights about energy transitions. Progress would be best served if critical social acceptance researchers were to bring their own creativity and ways of thinking to quantitative methods, where appropriate, rather than automatically rejecting them as incompatible.

Neither should strict rules be placed on the kinds of qualitative analyses considered valid in such a multidisciplinary problem space (Batel, 2018; Sandelowski, 2001). We are not exclusively quantitative researchers. We use qualitative or quantitative methods as the questions and populations demand. This frequently involves mixed methods, sometimes sequentially developing surveys to test interview-driven insight or conducting interviews to test the patterns seen at larger scales or in parallel looking towards independent triangulation (Walker & Baxter, 2019). That these elements are often published in different outlets is a practical matter of journal architecture, making more work for those who seek the whole picture rather than selected lenses. Simply put, however, RET acceptability calls for insight from many different minds and the methods that suit them, and diversity rather than orthodoxies.

That said, there are many topics of interest to critical social acceptance researchers for which quantitative methods like surveys will likely remain ill-suited. For instance, while we can ask respondents about their current contexts, it is challenging to explore their past ones, if any. If a person grew up in a coal mining town (Price, 2018) or on a hydroelectric reservoir (Sherren, Beckley, et al., 2016), for instance, what might that mean for their receptiveness to various energy sources? Structuring such a question to work for any survey respondent would be difficult, but may be possible if the focus of the survey is on one RET, which would make it more feasible to include additional questions to explore not only current but past experiences. We do, for instance, tend to ask respondents how long they have lived in the place they currently do. For relatively small geographies like the region we studied, follow-up archival research can help us understand what that period of occupancy may mean for their landscape experience and norms. But many other important experiences will remain invisible to us unless we go talk to people who lived or are living them. Those conversations will also provide a venue for rich explorations of many of the variables discussed here, such as trust, norms, preferences and emotions.

Finally, from a public policy perspective, local and regional governments are often interested in how their citizens will respond to new projects such as RET and associated landscape changes. Conventional opinion polling can offer some insight, but the methods described in this chapter offer an arguably better way to understand social acceptability. Building from these methods, we can identify scenarios, measure trade-offs and gain deeper insights into the norms, values and emotions associated with social acceptability. In the end, quantitative methods offer a way to explore public perceptions and test policy alternatives at the population level and thus have an important role to play in sustainability transitions.

References

Auspurg, K., & Hinz, T. (2014). *Factorial survey experiments* (Vol. 175). Sage.

Batel, S. (2018). A critical discussion of research on the social acceptance of renewable energy generation and associated infrastructures and an agenda for the future. *Journal of Environmental Policy & Planning, 20*(3), 356–369. https://doi.org/10.1080/1523908X.2017.1417120

Batel, S., & Devine-Wright, P. (2015). A critical and empirical analysis of the national-local 'gap' in public responses to large-scale energy infrastructures. *Journal of Environmental Planning and Management, 58*(6), 1076–1095. https://doi.org/10.1080/09640568.2014.914020

Batel, S., Devine-Wright, P., & Tangeland, T. (2013). Social acceptance of low carbon energy and associated infrastructures: A critical discussion. *Energy Policy, 58*, 1–5. https://doi.org/10.1016/j.enpol.2013.03.018

Bericat, E. (2016). The sociology of emotions: Four decades of progress. *Current Sociology, 64*(3), 491–513. https://doi.org/10.1177/0011392115588355

Bicchieri, C. (2017). *Norms in the wild: How to diagnose, measure, and change social norms*. Oxford University Press.

Biel, A., & Thøgersen, J. (2007). Activation of social norms in social dilemmas: A review of the evidence and reflections on the implications for environmental behaviour. *Journal of Economic Psychology, 28*(1), 93–112.

Camerer, C. F., Dreber, A., Holzmeister, F., Ho, T.-H., Huber, J., Johannesson, M., Kirchler, M., Nave, G., Nosek, B. A., Pfeiffer, T., Altmejd, A., Buttrick, N., Chan, T., Chen, Y., Forsell, E., Gampa, A., Heikensten, E., Hummer, L., Imai, T., ... Wu, H. (2018). Evaluating the replicability of social science experiments in Nature and Science between 2010 and 2015. *Nature Human Behaviour, 2*(9), 637–644. https://doi.org/10.1038/s41562-018-0399-z

Chan, K. M. A., Balvanera, P., Benessaiah, K., Chapman, M., Díaz, S., Gómez-Baggethun, E., Gould, R., Hannahs, N., Jax, K., Klain, S., Luck, G. W., Martín-López, B., Muraca, B., Norton, B., Ott, K., Pascual, U., Satterfield, T., Tadaki, M., Taggart, J., & Turner, N. (2016). Opinion: Why protect nature? Rethinking values and the environment. *Proceedings of the National Academy of Sciences, 113*(6), 1462–1465. https://doi.org/10.1073/pnas.1525002113

Chappell, E., Parkins, J. R., & Sherren, K. (2020). Climax thinking, place attachment, and utilitarian landscapes: Implications for wind energy development. *Landscape and Urban Planning, 199*, 103802. https://doi.org/10.1016/j.landurbplan.2020.103802

Chappell, E. N., Parkins, J. R., & Sherren, K. (in press). Those who support wind development in view of their home take responsibility for their energy use and that of others: evidence from a multi-scale analysis. *Journal of Environmental Policy & Planning*, 18pp. https://doi.org/10.1080/1523908X.2021.1888699

Clarke, C. E., Bugden, D., Hart, P. S., Stedman, R. C., Jacquet, J. B., Evensen, D. T. N., & Boudet, H. S. (2016). How geographic distance and political ideology interact to influence public perception of unconventional oil/natural gas development. *Energy Policy, 97*, 301–309. https://doi.org/10.1016/j.enpol.2016.07.032

Daggett, C. (2018). Petro-masculinity: Fossil fuels and authoritarian desire. *Millennium, 47*(1), 25–44. https://doi.org/10.1177/0305829818775817

Darnhofer, I., Lamine, C., Strauss, A., & Navarrete, M. (2016). The resilience of family farms: Towards a relational approach. *Journal of Rural Studies, 44*, 111–122. https://doi.org/10.1016/j.jrurstud.2016.01.013

Devine-Wright, P. (2011). Public engagement with large-scale renewable energy technologies: Breaking the cycle of NIMBYism. *WIREs Climate Change, 2*(1), 19–26. https://doi.org/10.1002/wcc.89

Dryzek, J. S., & Berejikian, J. (1993). Reconstructive democratic theory. *American Political Science Review, 87*(1), 48–60. https://doi.org/10.2307/2938955

Ehrlich, P. R., & Levin, S. A. (2005). The evolution of norms. *PLoS Biology, 3*(6), e194. https://doi.org/10.1371/journal.pbio.0030194

Ellis, G., Barry, J., & Robinson, C. (2007). Many ways to say 'no', different ways to say 'yes': Applying Q-Methodology to understand public acceptance of wind farm proposals. *Journal of Environmental Planning and Management, 50*(4), 517–551. https://doi.org/10.1080/09640560701402075

Farrow, K., Grolleau, G., & Ibanez, L. (2017). Social norms and pro-environmental behavior: A review of the evidence. *Ecological Economics, 140*, 1–13. https://doi.org/10.1016/j.ecolecon.2017.04.017

Fergen, J., & Jacquet, J. (2016). Beauty in motion: Expectations, attitudes, and values of wind energy development in the rural U.S. *Energy Research & Social Science, 11*, 133–141. https://doi.org/10.1016/j.erss.2015.09.003

Firestone, J., Bates, A., & Knapp, L. A. (2015). See me, Feel me, Touch me, Heal me: Wind turbines, culture, landscapes, and sound impressions. *Land Use Policy, 46*, 241–249. https://doi.org/10.1016/j.landusepol.2015.02.015

Firestone, W. A. (1987). Meaning in method: The rhetoric of quantitative and qualitative research. *Educational Researcher, 16*(7), 16–21. https://doi.org/10.3102/0013189X016007016

Giddens, A. (1979). *Central problems in social theory: Action, structure and contradiction in social analysis.* Macmillan Press.

Goodale, K., Parsons, G., & Sherren, K. (2015). The nature of the nuisance—damage or threat—determines how perceived monetary costs and cultural benefits influence farmer tolerance of wildlife. *Diversity, 7*(3), 318–341. https://doi.org/10.3390/d7030318

Gross, C. (2007). Community perspectives of wind energy in Australia: The application of a justice and community fairness framework to increase social acceptance. *Energy Policy, 35*(5), 2727–2736. https://doi.org/10.1016/j.enpol.2006.12.013

Grubert, E. (2018). Relational values in environmental assessment: The social context of environmental impact. *Current Opinion in Environmental Sustainability, 35*, 100–107. https://doi.org/10.1016/j.cosust.2018.10.020

Hall, N., Ashworth, P., & Devine-Wright, P. (2013). Societal acceptance of wind farms: Analysis of four common themes across Australian case studies. *Energy Policy, 58*, 200–208. https://doi.org/10.1016/j.enpol.2013.03.009

Holowach, M. C., Parkins, J., Anders, S. M., & Meyerhoff, J. (2019). *Ownership structure, fairness and the future of wind energy development in Canada – A survey-experimental study.* Retrieved December 28, 2019, from https://ageconsearch.umn.edu/record/291103/files/Abstracts_19_05_15_01_37_10_91__98_155_87_27_0.pdf

Iftekhar, M. S., & Pannell, D. J. (2015). "Biases" in adaptive natural resource management. *Conservation Letters, 8*(6), 388–396. https://doi.org/10.1111/conl.12189

Liebe, U., Bartczak, A., & Meyerhoff, J. (2017). A turbine is not only a turbine: The role of social context and fairness characteristics for the local acceptance of wind power. *Energy Policy, 107*, 300–308. https://doi.org/10.1016/j.enpol.2017.04.043

Maehr, A. M., Watts, G. R., Hanratty, J., & Talmi, D. (2015). Emotional response to images of wind turbines: A psychophysiological study of their visual impact on the landscape. *Landscape and Urban Planning, 142*, 71–79. https://doi.org/10.1016/j.landurbplan.2015.05.011

Mann, C., Parkins, J. R., Isaac, M. E., & Sherren, K. (2019). Do practitioners of holistic management exhibit systems thinking? *Ecology and Society, 24*(3), 19. https://doi.org/10.5751/ES-11092-240319

Morrow, R. A., & Brown, D. D. (1994). *Critical theory and methodology* (Vol. 3). Sage.

Paluck, E. (2009). What's in a norm? Sources and processes of norm change. *Journal of Personality and Social Psychology, 96*(3), 594–600. https://doi.org/10.1037/a0014688

Parkins, J. R., Beckley, T., Comeau, L., Stedman, R. C., Rollins, C. L., & Kessler, A. (2017). Can distrust enhance public engagement? Insights from a national survey on energy issues in Canada. *Society & Natural Resources, 30*(8), 934–948. https://doi.org/10.1080/08941920.2017.1283076

Parkins, J. R., Chappell, E., & Sherren, K. (2020). *The paradox of female support for coal power: Insights on knowledge, values and trust from a survey of Canadians.* Presented online at the International Symposium of Society and Resource Management, July 2020.

Parkins, J. R., Hempel, C., Beckley, T. M., Stedman, R. C., & Sherren, K. (2015). Identifying energy discourses in Canada with Q methodology: Moving beyond the environment versus economy debates. *Environmental Sociology, 1*(4), 304–314. https://doi.org/10.1080/23251042.2015.1054016

Parkins, J. R., Rollins, C., Anders, S., & Comeau, L. (2018). Predicting intention to adopt solar technology in Canada: The role of knowledge, public engagement, and visibility. *Energy Policy, 114*, 114–122. https://doi.org/10.1016/j.enpol.2017.11.050

Parkins, J. R., & Sherren, K. (2021). Identifying energy discourses across scales in Canada with Q methodology and survey research. In J. B. Jacquet, J. H. Haggerty, & G. L. Theodori (Eds.), *Energy impacts: A multidisciplinary exploration of North American energy development.* Social Ecology Press and University of Utah Press.

Price, D. (2018). *Coal cultures: Picturing mining landscapes and communities.* Bloomsbury Publishing.

Rand, J., & Hoen, B. (2017). Thirty years of North American wind energy acceptance research: What have we learned? *Energy Research & Social Science, 29*, 135–148. https://doi.org/10.1016/j.erss.2017.05.019

Sandelowski, M. (2001). Real qualitative researchers do not count: The use of numbers in qualitative research. *Research in Nursing & Health, 24*(3), 230–240. https://doi.org/10.1002/nur.1025

Schlüter, M., Baeza, A., Dressler, G., Frank, K., Groeneveld, J., Jager, W., Janssen, M. A., McAllister, R. R. J., Müller, B., Orach, K., Schwarz, N., & Wijermans, N. (2017). A framework for mapping and comparing behavioural theories in models of social-ecological systems. *Ecological Economics, 131*, 21–35. https://doi.org/10.1016/j.ecolecon.2016.08.008

Sherren, K., Beckley, T., Greenland-Smith, S., & Comeau, L. (2017). How provincial and local discourses aligned against the prospect of dam removal in New Brunswick, Canada. *Water Alternatives, 10*(3), 697. http://www.water-alternatives.org/index.php/alldoc/articles/vol10/v10issue3/378-a10-3-4/file

Sherren, K., Beckley, T. M., Parkins, J. R., Stedman, R. C., Keilty, K., & Morin, I. (2016). Learning (or living) to love the landscapes of hydroelectricity in Canada: Eliciting local perspectives on the Mactaquac Dam via headpond boat tours. *Energy Research & Social Science, 14*, 102–110. https://doi.org/10.1016/j.erss.2016.02.003

Sherren, K., Loik, L., & Debner, J. A. (2016). Climate adaptation in 'new world' cultural landscapes: The case of Bay of Fundy agricultural dykelands (Nova Scotia, Canada). *Land Use Policy, 51*, 267–280. https://doi.org/10.1016/j.landusepol.2015.11.018

Sherren, K., Parkins, J. R., Owen, T., & Terashima, M. (2019). Does noticing energy infrastructure influence public support for energy development? Evidence from a national survey in Canada. *Energy Research & Social Science, 51*, 176–186. https://doi.org/10.1016/j.erss.2019.01.014

Stern, M. J. (2018). *Social science theory for environmental sustainability: A practical guide*. Oxford University Press.

Stern, M. J., & Baird, T. D. (2015). Trust ecology and the resilience of natural resource management institutions. *Ecology and Society, 20*(2) http://www.jstor.org/stable/26270214

Stern, M. J., & Coleman, K. J. (2015). The multidimensionality of trust: Applications in collaborative natural resource management. *Society & Natural Resources, 28*(2), 117–132. https://doi.org/10.1080/08941920.2014.945062

Walker, C., & Baxter, J. (2017). Procedural justice in Canadian wind energy development: A comparison of community-based and technocratic siting processes. *Energy Research & Social Science, 29*, 160–169. https://doi.org/10.1016/j.erss.2017.05.016

Walker, C., & Baxter, J. (2019). Method sequence and dominance in mixed methods research: A case study of the social acceptance of wind energy literature. *International Journal of Qualitative Methods, 18*, 1609406919834379. https://doi.org/10.1177/1609406919834379

Walker, C., Baxter, J., & Ouellette, D. (2014). Beyond rhetoric to understanding determinants of wind turbine support and conflict in two Ontario, Canada communities. *Environment and Planning A: Economy and Space, 46*(3), 730–745. https://doi.org/10.1068/a130004p

Wicherts, J. M., Veldkamp, C. L., Augusteijn, H. E., Bakker, M., Van Aert, R., & Van Assen, M. A. (2016). Degrees of freedom in planning, running, analyzing, and reporting psychological studies: A checklist to avoid p-hacking. *Frontiers in Psychology, 7*, 1832. https://doi.org/10.3389/fpsyg.2016.01832

Zaunbrecher, B., Arning, K., & Ziefle, M. (2018). *The good, the bad and the ugly: Affect and its Role for renewable energy acceptance*. Proceedings of the 7th International Conference on Smart Cities and Green ICT Systems (SMARTGREENS 2018), Madeira, Portugal.

CHAPTER 3

Using a Critical Approach to Unpack the Visual-Spatial Impacts of Energy Infrastructures

Susana Batel and Patrick Devine-Wright

What Is in the Visual Impact of Renewable Energy and Associated Technologies?

Research on the social acceptance of renewable energy technologies (RET) has pointed out how several factors contribute to local opposition to RET (Ellis & Ferraro, 2016; Wüstenhagen et al., 2007). One of those factors, and arguably the one which has received more attention in the literature, refers to the so-called visual impacts of RET. However, as Wolsink (2018) recently highlighted, "visual impact is conceptually a highly complex and frequently misunderstood topic in studies on RES[T]" (p. 552; see also

S. Batel (✉)
Instituto Universitário de Lisboa (ISCTE-IUL), Cis-IUL, Lisboa, Portugal
e-mail: susana.batel@iscte-iul.pt

P. Devine-Wright
University of Exeter, Exeter, UK
e-mail: p.g.devine-wright@exeter.ac.uk

S. Batel, D. Rudolph (eds.), *A critical approach to the social acceptance of renewable energy infrastructures*,
https://doi.org/10.1007/978-3-030-73699-6_3

Kropp, 2018). Throughout the years, social science research on the social acceptance of RET has used concepts such as 'visual impact', 'physical appearance', 'project-place fit' and 'impacts on landscape'. These are often used interchangeably, without a clear definition of what they mean exactly and if and how they might differ, with 'visual impact' particularly being used as a shorthand to talk about any relation between RET and space (for some illustrations see Cowell, 2010, p. 223; Devine-Wright & Howes, 2010, p. 272; Naumann & Rudolph, 2020, p. 100; Wolsink, 2000, p. 51). In this paper, we build upon recent calls for a better understanding of the relations between infrastructures, space, landscapes and the rural in energy transitions (Jefferson, 2018; Naumann & Rudolph, 2020; Wolsink, 2018), in order to critically review the literature in this regard.

When the social sciences began to take an interest in understanding local opposition to RET, this was normally explained by the NIMBY (Not In My Backyard) phenomenon (Wolsink, 2000). NIMBY was based on a 'backyard' representation of the (rural) contexts where RET are normally deployed and that equated the proximity and visibility of RET with opposition (Devine-Wright, 2011a). Within this approach to local opposition, rejection of RET was therefore mostly due to individuals' cognitive perceptions of wind farms and other large scale energy infrastructures as lacking certain visual qualities and coherence in relation to the surrounding context/landscape[1] (e.g. Sevenant & Antrop, 2010; Tveit et al., 2006). As Wolsink (2018) highlights, this has created the misunderstanding, especially among policy-makers and developers, that "common-sense 'knowledge' simply associates wind turbines and transmission lines with negative visual attitudes ('visual pollution'), as evidenced in many studies applying economic valuation approaches (willingness-to-pay) for visual impact" (p. 552; e.g. Ladenburg & Dubgaard, 2007). As suggested by Wolsink (2018), this has had the consequence not only of equating visual impact with mere visibility, but also of considering visual impact as only materialised in the physical appearance of turbines and pylons, such as their size, colour and shape. Within this perspective, the visual dimension of RET and its impact on people's responses can then be seen mainly as being conceived as an individual and personal process, a matter of individuals' idiosyncratic aesthetic preferences and perceptions (see Lothian, 2008; Sibille et al., 2009), and local opposition as mainly stemming from what object is being deployed, with what aesthetic/visual characteristics. As such, it is a perspective that commodifies the relation between RET and their spatial deployment. This means that by focusing mainly on RET's physical appearance and visibility aspects, this perspective conceives them as a product, as something that individuals—or consumers (Lennon et al., 2019)—might like less or more, and as therefore amenable to economic valuation (Wolsink, 2018) within neoliberal capitalist political and economic systems.

As research on social acceptance of RET increased, so did critiques of the NIMBY explanation to local opposition and, with it, calls to overcome more individual-centred approaches to understanding local opposition (see Burningham, 2000; Ellis et al., 2007 for critiques). One of the most prolific alternative theories to NIMBY has been the so-called place attachment perspective (Devine-Wright, 2009). This perspective proposes that people's emotional and symbolic relations with the place where they live will impact on their acceptance, rejection or ambivalence towards RET in their locality depending on how these RET are seen as fitting or not that place (McLachlan, 2009). For instance, Devine-Wright and Howes (2010) and Devine-Wright and Devine-Wright (2009) have shown how in different locations in the UK and regarding wind farms and high voltage power lines respectively, people's responses to these RET were either mainly positive or mainly negative depending on the symbolic meanings and emotional connections that they had in relation to the places where those RET were going to be deployed and how the RET projects were seen as fitting those meanings and connections or instead menacing them. Something which this body of research illustrates is that what has often been deemed as the visual impact of RET and individuals' preferences for the physical appearance of certain RET's instead of others, is actually not an individual or personal, cognitive and perceptual process. Instead, it is part of an ongoing spatial process of people-place relations at a local, community level, which also implicates meaning-making and emotional attachments.

Along with critiques of NIMBY that focus on the role of local place dimensions, others have pinpointed the role played by more cultural and institutional factors in local opposition, such as the role of landscape traditions and their reproduction in land-use planning systems (Cowell, 2010; Woods, 2005), different mass media (Batel, 2020; Eriksson, 2010; Phillips et al., 2001) and everyday practices and discourses, such as those of rural inhabitants and 'users' (Batel et al., 2015; Wallwork & Dixon, 2004). This research has been concerned with representations of landscapes as national/cultural/societal/ideological level imaginaries (Gailing & Leibenath, 2015; Kropp, 2018), practices and relations. Specifically, whether these have changed or not over time, if and how they might differ across cultural-spatial-geographical contexts (Gkartzios & Remoundou, 2018; Phadke, 2011; Toke et al., 2008) and how, in turn, those changes and differences might account for local opposition. For instance, a large amount of this research has suggested how local opposition to RET in the UK can be largely explained by the impact of representations of the British countryside as natural, pristine, non-human, 'our green and pleasant land', on people's ideas and emotions about RET being deployed in that landscape. In turn, this hegemonic representation of rural landscapes in the UK has been more recently picked up by urban to rural migrants to justify and legitimate their appropriation of rural areas. It underpins their opposition to the change of that countryside idyll, or, as discussed by the social sciences, a "consuming idyll", part of a broader process of commodification of the rural,[2] also described as "McRuralisation" (Eriksson, 2010; Jansson, 2013; Silva, 2013).

However, this consuming idyll of the countryside often coexists with representations of the countryside as productive, as the place for farming and making a living. This representation of rural productivism might then be amenable to "a reversal of the post-productive discourse, incorporating rural areas into new production and capital accumulation processes" (Naumann & Rudolph, 2020, p. 100) in the guise of RET being deployed in rural landscapes (also Silva & Prista, 2016). In fact, and as clearly put by Phadke (2011), "wind energy opposition politics are essentially battles over rural space; over who controls the productive and consumptive qualities of rural landscapes" (p. 756).

In sum, we can consider that so far the relation between RET, space and local opposition has been loosely referred to as 'the visual impacts of RET', even if this label actually translates three different, but interrelated, ways through which the relations between RET and space have been examined so far: first, with a focus on individual preferences for the physical characteristics and impacts of RET; second, with a focus on people-place symbolic and emotional 'fit' at a local, community level; third, with a focus on landscape traditions as cultural, institutional and ideological representations. In this chapter, we argue that research needs to more explicitly recognise these levels and their interrelations or, in other words, to

- adopt a more dynamic and relational approach (Batel, 2018; Labussière & Nadaï, 2018) to the rural visual-spatial dimension of the deployment of RET that considers the interrelations between individual, local/community and cultural levels, as illustrated in Fig. 3.1;

- overcome conceptions of individuals' ideas about the physical appearance of RET as solely cognitive/perceptual and re-present them instead as positional, as the articulation of a combination of cultural and local representations of landscape and place, and more subjective representations, based on life-place experiences and trajectories (see Bailey et al., 2016) and associated political-ideological dimensions;[3]
- adopt conceptual and analytical tools to uncover the individual, local/community and cultural/institutional dimensions of people's representations about the visual-spatial impacts of RET, as distinct but closely interrelated dimensions of a broader system of representation and meaning-making.

Building on these dimensions, we now direct attention to the social-psychological processes of anchoring and objectification defined and explored by the theory of social representations (TSR; Batel & Devine-Wright, 2015; Moscovici, 1961/72). These will help us unpack and identify the dimensions of the visual-spatial impact of RET in discourse and communication, in people's meaning-making and how to examine them for a better understanding of the relation between RET's visual-spatial impact and people's responses to RET.

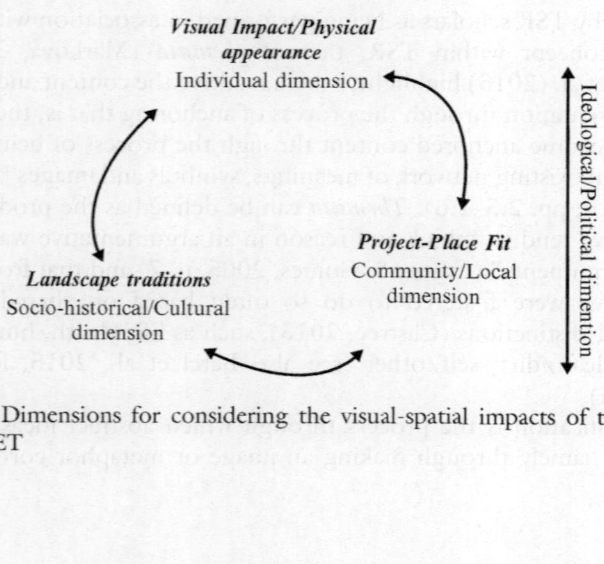

Fig. 3.1 Dimensions for considering the visual-spatial impacts of the deployment of RET

Anchoring and Objectification

The theory of social representations (TSR; Batel & Devine-Wright, 2015; Howarth, 2006; Moscovici, 1961/72) aims to understand how meanings are constructed, negotiated and contested by people through discourse and communication, therefore promoting or resisting social change. One of the main domains to which TSR has been applied has been that of cultural and techno-scientific change—such as RET (e.g. Batel & Devine-Wright, 2015; Devine-Wright & Devine-Wright, 2009)—and how that is appropriated in contemporary public spheres. As such, the theory has been seen as very well equipped to look at the extraordinary, or at moments when social change is just introduced into a society or in a community, again, just like with RET. In other words then, it is a socio-psychological theory about how the unfamiliar becomes familiar (or not) over time and in specific socio-spatial-cultural contexts (Moscovici, 1961/72; Wagner, 1998). Anchoring and objectification have been proposed as precisely the two processes[4] through which people attempt to turn the unfamiliar into familiar knowledge,[5] ideas and practices. Anchoring has been described as allowing the classification of new social objects or the unfamiliar into previous and familiar knowledge that makes up our cultures and societies, and is often done through processes such as naming. As exemplified by Höijer (2017), this might involve that "a new ill-health is called the Black Death of our age, the complex scientific phenomenon climate change may be shortly labelled as the weather and so on" (p. 7). Anchoring has also been identified by TSR scholars as being performed in association with another relevant concept within TSR, that of *themata* (Marková, 2003). As Moloney et al. (2015) highlight, "themata drive the content and structure of a representation through the process of anchoring that is, themata take form or become anchored content through the process of being embedded into an existing network of meanings, symbols and images" (Moloney et al., 2015, pp. 2.5–2.6). *Themata* can be defined as the product of the fact that we tend to "think and reason in an argumentative way, by contrasting arguments" (Castro & Gomes, 2005, p. 7) and that from modernity on we were fostered to do so often based on shared symbolic essentialist distinctions (Castree, 2013), such as life/death, human/non-human, clean/dirt, self/other (see also Batel et al., 2015, for further discussion).

Objectification is the process through which abstract ideas are made concrete, namely through making an image or metaphor correspond to the object. As suggested by Wagner and Kronberger (2001), "it has an image structure that visibly reproduces a complex of ideas and weaves it into the fabric of the group's common sense" (p. 5; for an example on high voltage power lines see Devine-Wright & Devine-Wright, 2009). As these definitions suggest, it is not always easy to distinguish anchoring from objectification—they should be seen as interrelated or different parts of a continuum (Wagner & Kronberger, 2001), with anchoring more based in comparing one object with another and integrating new objects into familiar elements, and objectification more based in summarising abstract ideas into a simple(r) image.

Anchoring and objectification might then be useful analytical tools to better understand how the three dimensions (individual, local/community and socio-cultural) shape the relation between the visual-spatial impact of RET and people's responses to them. We will now turn to examining data materials to illustrate that.

Illustrating Anchoring and Objectification of the Visual-Spatial Impact of RET and People's Responses

To illustrate how anchoring and objectification might be helpful analytical tools to identify and better understand the role of the visual-spatial impact at local and cultural/institutional levels, and people's responses to RET, we will refer to examples and extracts from qualitative data collected in British newspapers. Articles were collected from the UK press (Guardian, Times and Sun newspapers) in the period 2008–2014 (for more information on this dataset, see Batel, 2020).

Visual-Spatial Impact as Landscape Traditions

As discussed before, anchoring has been described as allowing the classification of new social objects or the unfamiliar into previous and familiar knowledge that makes up our societies, cultures and traditions, often in the guise of *themata*, with objectification summarising that process, mainly through images. This is quite clear in how the impacts of RET on landscape traditions are anchored in familiar, socio-historically and culturally relevant landmarks and objects and objectified in societal relevant tropes—such as monsters—often with the help of socio-historically significant *themata*. This is clear in Extract 3.1.

Extract 3.1

a. *erecting [wind] turbines as tall as the Gherkin office tower in* ***the City of London around the coast of Britain*** [Times, 833:2];
b. *wind turbines have gone from being perceived as obtrusive bits of heavy machinery to quite elegant architecture* [Times, 763: 10];
c. *These Round 3 wind farms will operate at depths never tried before and far from shore, using unproven* ***giant*** *turbines twice as tall as Big Ben* [Times, 667:4]
d. *Five turbines, each more than 400ft tall—almost the height of Blackpool Tower* [Times, 756:5]
e. *It will be like looking through the bars of a prison and the Scottish citizens will be the prisoners* [Times, 686:9]

Something in common across the extracts presented above is that they discuss wind farms not in terms of specific aesthetic characteristics and preferences nor in terms of the fit of a particular wind farm project with a particular place, but instead how wind farms in general are made sense of in this day and age. Anchoring as a process is illustrated in Extracts 3.1a–3.1e when wind farms are appropriated and made sense of by comparing them with contemporary and familiar landmarks and objects, such as the Gherkin office tower and Big Ben in London (Extracts 3.1a and 3.1b), Blackpool Tower in Lancashire (Extract 3.1d) and the bars of a prison (Extract 3.1e). The comparison with Blackpool Tower seems to be quite culturally significant, given that it has been identified by previous research as an anchoring object to wind farms in the UK (see Devine-Wright & Howes, 2010, p. 275). Blackpool Tower is a 158-m-high tower in Blackpool, Lancashire, UK, that was constructed during the 'British Empire' as the tallest human-made structure. It is relevant then how this is one of the most common referents to anchor wind farms in terms of their size and scale, as a way to accentuate their human-made, industrial and unnatural characteristics, and thus their incompatibility with how rural landscapes are represented in the current socio-historical period (see also Extract 3.2), as (in the UK) mainly 'flat' spaces where oak trees are the tallest natural phenomena. However, different socio-historical, geographical and cultural contexts could dictate otherwise, as exemplified by

Furby et al. (1988) regarding high voltage power lines in the USA, where, during the 1950–1960s their deployment in the North American countryside was seen instead as a sign of progress (see also Bosch & Schmidt, 2020). Extract 3.1 illustrates in a clear way how wind farms in the UK are nowadays instead constructed based on *themata* of human/non-human, natural/unnatural/industrial, rural/urban and destruction/progress (also very evident in Extract 3.1b), based on their size, scale and shape, through which they are objectified as giant and monstrous and, as such, rejected. This is better illustrated in Extract 3.2.

Extract 3.2

a. *Abington is in South Lanarkshire and the countryside there is a forgotten jewel. It is not designated, so there is an open season for industrialization* [Times, 764:18]
b. *As conservationists battle to slow or even stop the relentless march of wind turbines across the countryside* [Times, 974: 2]
c. *Living in a country with a landscape blighted by 300,000 vacant homes has taught us one thing: progress really can spoil the view* [Times, 1008:7]

This set of extracts illustrates that anchoring and objectifying wind farms as oversized, giant, as tall as or even bigger than some of the tallest landmarks in the UK, is closely related with making sense of them with the help of culturally and historically significant *themata* such as the ones just mentioned before and others such as peace/war (Extract 3.2b) and healthy/unhealthy (Extract 3.3c). These are used by people to make sense of wind farms and their impact on the British rural landscape at the same time as allowing them to politically position themselves in relation to that (Batel et al., 2015). In other words, and as already briefly discussed in the introduction of this chapter, while positing the countryside as natural and pristine, as an idyll to be preserved, and wind farms as industrial, a disease and the enemy, these ways of anchoring and objectifying wind farms in relation to the rural and associated landscapes also reflect ideological underpinning by further commodifying these spaces as places for the consumption of a certain (aesthetic and leisure) version of the rural and landscapes by certain groups of people, such as urban to rural migrants and upper classes (Eriksson, 2010; Gkartzios & Remoundou, 2018; Silva, 2013).

Visual-Spatial Impact as Project-Place Fit

Extract 3.1a also talks about how people's sense-making of the visual-spatial impacts of wind farms is related not only with when that is happening—the socio-historical/ideological period—but also with where. In that extract, the fact that wind farms are as tall as the Gherkin Tower which is in the City of London—in a commercial area within a very urbanised city—and that the wind farms are going to be deployed around the coast of Britain, is highlighted, emphasising the lack of fit between building something so tall in an urban area and in a rural, coastal area. This is also clear in Extract 3.3.

Extract 3.3

a. *their astonishment that anyone could be so mad, so vandalistic as to site* ***these monstrosities*** *in such a way as to interfere with the view of Bow from Dartmoor* [in the South West of England], *or Dartmoor from Bow* [Times, 728:9]
b. *Above the braes that line the coast around nearby Luce Bay a different landscape can be found, featuring the highest point in the Machars—Mochrum Fell* [in Scotland]. *Here, in fields enclosed by centuries old dry-stone dykes, Galloway cattle and sheep graze (…) Mochrum Fell will be* ***dwarfed*** *by the* ***towering steel giants*** *that on clear days will be visible from Northern Ireland* [Times, 820:12]
c. *One of the town's councillors wrote: "The Ardrossan wind farm has been overwhelmingly accepted by local people—instead of spoiling the* [Scottish] *landscape, we believe it has been enhanced. The turbines are impressive looking, bring a calming effect to the town and, contrary to the believe that they would be noisy, we have found them to be* ***silent workhorses****"* [The Guardian, P339]

In Extract 3.3, and similarly to Extract 3.2a, discourses not only represent wind farms in general and in relation to British landscapes more generally, but also discuss specific wind farms to be built in particular places/communities. As such, people discuss particular attachments to specific views and places that will be affected by the construction of those 'giant monstrosities' (Extracts 3.3a and 3.3b). Interestingly, Extract 3.3c discusses this RET-place fit and related symbolic and emotional connections in a positive way. It highlights how the local community has accepted the

Ardrossan wind farm based on the belief that the local landscape is going to be enhanced and not spoiled and objectifying the wind turbines as *silent workhorses* instead of as 'monstrosities' and 'giants'. This way of objectifying the wind farm is particularly interesting because it highlights how objectification is not necessarily negative or oppositional (see also Devine-Wright, 2011b). It can also be employed to make a connection with rural areas' historic farming and productivist functions, providing a sense of continuity between past agricultural practices of land farming and present-day energy practices of wind farming (see also Naumann & Rudolph, 2020).

Visual-Spatial Impact as Physical Appearance

The extracts presented so far illustrate how people make sense of wind farms in relation to both specific landscape and associated ideological traditions, and more local places through anchoring them in familiar landmarks, objects and experiences, and objectifying them in familiar societal tropes to deal with what is often considered threatening, strange and unwanted, such as 'monsters' or 'giants'. These extracts clearly reveal that these meaning-making processes are deeply political, as they put forward and reproduce specific versions of the British rural countryside instead of others, and also, in an associated way, of certain energy landscapes instead of others (Bosch & Schmidt, 2020). In turn, making sense of the spatial impact of RET at those two levels impacts on how people, as individuals, make sense of wind farms as physical and material objects, to either consider them as beautiful (Extract 3.4b) or instead to try to minimise their visual impact (Extract 3.4a), which has been common practice as a mitigation measure by wind energy companies in recent years.

Extract 3.4

a. *she uses techniques such as painting the turbines in hues to match the average tone of the sky, to minimize their visual impact* [Guardian, 763:15]
b. *But visual impact is clearly Berkeley's* [a small British town] *main concern.* ***For some, wherever turbines are built will always be inappropriate***. *The fact that since I first saw them in Cornwall some 15 years ago I have found them beautiful is neither here nor there—but it is relevant that repeated opinion surveys have shown that more than 70% of people in this country either like or do not mind the visual impact* [Guardian, 339:2]

c. *Opponents are appalled by the size of the turbines, which stand 300ft (91m) high* [Times, 764:11]

These extracts however illustrate not only how RET are represented as objects but also how that is shaped by the context where they are located, or, in other words, how spatial impact as visual and physical appearance is rooted in how RET are perceived both generally as particular technological artefacts within a certain socio-cultural and historical context (Extract 3.4b: *more than 70% of people in this country either like or do not mind the visual impact*) and as particular infrastructures to be built in particular places (Extract 3.4b: *the fact that since I first saw them in Cornwall some 15 years ago I have found them beautiful*). They also make us consider that perhaps a more fruitful and appropriate way to develop research on this more positional or personal dimension of the relation between the visual impact of RET and people's responses to that would be to, instead of adopting the traditional cognitivist, perceptual approach, adopt a phenomenological perspective focusing on lived, embodied experiences in place (e.g. Seamon, 2014). This perspective could explore how particular life-place trajectories (i.e. interpretative and emotional experiences and biographies, see Bailey et al., 2016) articulate cultural-institutional, contextual-local and ideological-political meanings[6] and relations to make sense of particular RET in particular places at a given socio-historical period.

Conclusions and Discussion

This chapter takes as its point of departure recent calls to further understand the visual impact of RET and their role in people's responses (Wolsink, 2018) and also for further considering the impact of RET on the rural and on its meanings (Naumann & Rudolph, 2020). In fact, research in this area so far has tended to mostly conceive the visual impact of RET from a positivist, individualist and cognitivist perspective, focused on how the visibility of specific infrastructures and their aesthetic characteristics would make them more or less acceptable from a personal point of view. In this chapter we propose instead that research in this area should acknowledge and further explore the visual-spatial impacts of RET and their relation with people's responses to RET based on three key interrelated dimensions: visual-spatial impacts as (a) landscape ideological traditions, (b) project-place fit and (c) physical appearance. In so doing, we also proposed that for analysing how people make sense of the visual-spatial impacts of RET at those three levels, it might be relevant to examine the socio-psychological processes of anchoring and objectification, and associated *themata*, as proposed by the theory of social representations. These processes reveal how people make sense of RET at different levels (personal/positional, community or group, socio-cultural) and how, for that, they rely on socio-culturally, historically and psychologically relevant landmarks, objects, stories, tropes, metaphors and images, which help them deal with—resist, support, tolerate—the change brought by RET to particular places.

As such, investigating people's meaning-making of the visual-spatial impact of RET through this three-fold lens, allows us to fully consider landscapes and the rural as "cosmopolitical 'things' that bring people together through their heterogeneous and their long-term social relevance" (Kropp, 2018, p. 563; also Naumann & Rudolph, 2020). It also allows to be more critical in research on the social acceptance of RET, namely by shining light on the power relations that shape whose versions of landscapes are more hegemonic and able to pervade societies—or the ideological dimensions of RET. On one hand, the commodification of the rural in the UK as the idyllic and pastoral British countryside to be protected from the giant monstrous wind farms and kept natural and pristine for its aesthetic consumption by, mostly, urban to rural migrants or visitors, is evident in these vocal (mainly negative) discourses of opposition to wind farms (see Bell et al., 2005). On the other hand, the so-called low-carbon energy transition—the move away from fossil fuels towards mainly large-scale renewable energy generation and associated infrastructures—is progressing and with it the commodification of the wind, sun and other common natural resources (Labussière & Nadaï, 2018) and what we could call the creative destruction of rural landscapes by RET within current neoliberal capitalist systems (Tonts & Greive, 2002; also Harvey, 2007), and related inequalities and dispossessions.

In both cases however, there is the danger of constraining the discussion and negotiation of alternative versions of RET, the rural and landscapes, that might give more voice and recognition to other groups and viewpoints—such as local and global ecosystems, farmers, migrant land workers, lower-income rural dwellers, tourists and other energy citizens—and, with it, of constraining more just energy transitions. In other words, this more multifaceted, relational way to look at the visual-spatial impact of RET allows us to further consider the complexity of people's responses to RET in the interrelation of these responses with the larger socio-economic, political, historical and cultural systems where they are embedded. As such, they highlight that these battles over the meaning of the countryside and the rural, while having very strong resonance in the UK as shown here, require cross-cultural research investigation across European states and worldwide.

Notes

1. Here we refer to landscape as the physical, visual and spatial materializations of certain geological, geographical and ecological characteristics that also reflect given socio-historical and political relations and practices and are in the making—or 'cosmopolitical' things, as put by Kropp (2018). Landscape is used in this chapter mainly to refer to these socio-physical settings in rural areas.
2. The rural should be seen as including not only the 'green and pleasant land' of the onshore countryside, but also other landscapes such as the coastline—see, for instance, Devine-Wright and Howes (2010), where the same dimensions were relevant to negotiate the deployment of offshore wind farms.
3. With ideological-political dimension we mean the value-laden and power-related processes that shape the societal/cultural, community/local and individual dimensions of the visual-spatial impacts of RET by reproducing, negotiating and/or contesting certain configurations of those impacts (see Woods, 1998; Wallwork & Dixon, 2004; Kropp, 2018).
4. Here conceived as socio-discursive and not as (socio-)cognitive.
5. Becoming familiar does not imply acceptance.
6. These include issues of justice and fairness as well in RET-related decision-making processes at national, regional and local levels, which can arguably also impact how people re-present the visual-spatial impacts of RET.

References

Bailey, E., Devine-Wright, P., & Batel, S. (2016). Using a narrative approach to understand place attachments and responses to power line proposals: The importance of life-place trajectories. *Journal of Environmental Psychology, 48*, 200–211. https://doi.org/10.1016/j.jenvp.2016.10.006

Batel, S. (2018). A critical discussion of research on the social acceptance of renewable energy generation and associated infrastructures and an agenda for the future. *Journal of Environmental Policy & Planning, 20*(3), 356–369. https://doi.org/10.1080/1523908X.2017.1417120

Batel, S. (2020). Re-presenting the rural in the UK press: An exploration of the construction, contestation and negotiation of media discourses on the rural within post-carbon energy transitions. *Energy Policy*. https://doi.org/10.1016/j.enpol.2020.111286

Batel, S., & Devine-Wright, P. (2015). Towards a better understanding of people's responses to renewable energy technologies: Insights from Social Representations Theory. *Public Understanding of Science, 24*(3), 311–325. https://doi.org/10.1177/0963662513514165

Batel, S., Devine-Wright, P., Wold, L., Egeland, H., Jacobsen, G., & Aas, O. (2015). The role of (de-)essentialisation within siting conflicts: An interdisciplinary approach. *Journal of Environmental Psychology, 44*, 149–159. https://doi.org/10.1016/j.jenvp.2015.10.004

Bell, D., Gray, T., & Haggett, C. (2005). The 'social gap' in wind farm siting decisions: Explanations and policy responses. *Environmental Politics, 14*(4), 460–477. https://doi.org/10.1080/09644010500175833

Bosch, S., & Schmidt, M. (2020). Wonderland of technology? How energy landscapes reveal inequalities and injustices of the German Energiewende. *Energy Research & Social Science, 70*, 101733. https://doi.org/10.1016/j.erss.2020.101733

Burningham, K. (2000). Using the language of NIMBY: A topic for research, not an activity for researchers. *Local Environment, 5*(1), 55–67. https://doi.org/10.1080/135498300113264

Castree, N. (2013). *Making sense of nature*. Routledge.

Castro, P., & Gomes, I. (2005). Genetically modified organisms in the Portuguese press: Thematization and anchoring. *Journal for the Theory of Social Behaviour, 35*(1), 1–17. https://doi.org/10.1111/j.0021-8308.2005.00261.x

Cowell, R. (2010). Wind power, landscape and strategic, spatial planning – The construction of 'acceptable locations' in Wales. *Land Use Policy, 27*(2), 222–232. https://doi.org/10.1016/j.landusepol.2009.01.006

Devine-Wright, H., & Devine-Wright, P. (2009). Social representations of electricity network technologies: Exploring processes of anchoring and objectification through the use of visual research methods. *British Journal of Social Psychology, 48*(2), 357–373. https://doi.org/10.1348/014466608X349504

Devine-Wright, P. (2009). Rethinking NIMBYism: The role of place attachment and place identity in explaining place-protective action. *Journal of Community & Applied Social Psychology, 19*(6), 426–441. https://doi.org/10.1002/casp.1004

Devine-Wright, P. (2011a). From backyards to places: Public engagement and the emplacement of renewable energy technologies. In P. Devine-Wright (Ed.), *Renewable energy and the public: From NIMBY to Participation (pp. 89-102)*. Routledge.

Devine-Wright, P. (2011b). Place attachment and public acceptance of renewable energy: A tidal energy case study. *Journal of Environmental Psychology, 31*, 336–343. https://doi.org/10.1016/j.jenvp.2011.07.001

Devine-Wright, P., & Howes, Y. (2010). Disruption to place attachment and the protection of restorative environments: A wind energy case study. *Journal of Environmental Psychology, 30*(3), 271–280. https://doi.org/10.1016/j.jenvp.2010.01.008

Ellis, G., Barry, J., & Robinson, C. (2007). Many ways to say 'no', different ways to say 'yes': Applying Q-methodology to understand public acceptance of wind farm proposals. *Journal of Environmental Planning and Management, 50*(4), 517–551. https://doi.org/10.1080/09640560701402075

Ellis, G., & Ferraro, G. (2016). *The social acceptance of wind energy. Where we stand and the path ahead.* JRC Science for Policy Report. European Commission, Brussels.

Eriksson, M. (2010). "People in Stockholm are smarter than countryside folks" – Reproducing urban and rural imaginaries in film and life. *Journal of Rural Studies, 26*(2), 95–104. https://doi.org/10.1016/j.jrurstud.2009.09.005

Furby, L., Slovic, P., Fischhoff, B., & Gregory, R. (1988). Public perceptions of electric power transmission lines. *Journal of Environmental Psychology, 8*(1), 19–43. https://doi.org/10.1016/S0272-4944(88)80021-5

Gailing, L., & Leibenath, M. (2015). The social construction of landscapes: Two theoretical lenses and their empirical applications. *Landscape Research, 40*(2), 123–138. https://doi.org/10.1080/01426397.2013.775233

Gkartzios, M., & Remoundou, K. (2018). Language struggles: Representations of the countryside and the city in an era of mobilities. *Geoforum, 93*, 1–10. https://doi.org/10.1016/j.geoforum.2018.04.016

Harvey, D. (2007). Neoliberalism as creative destruction. *The Annals of the American Academy of Political and Social Science, 610*(1), 21–44.

Höijer, B. (2017). Social Representations Theory: A new theory for media research. *Nordicom Review, 32*, 3–16. https://doi.org/10.1515/nor-2017-0109

Howarth, C. (2006). A social representation is not a quiet thing: Exploring the critical potential of social representations theory. *British Journal of Social Psychology, 45*(1), 65–86. https://doi.org/10.1348/014466605X43777

Jansson, A. (2013). The hegemony of the urban/rural divide: Cultural transformations and mediatized moral geographies in Sweden. *Space and Culture, 16*(1), 88–103. https://doi.org/10.1177/1206331212452816

Jefferson, M. (2018). Safeguarding rural landscapes in the new era of energy transition to a low carbon future. *Energy Research & Social Science, 37*, 191–197. https://doi.org/10.1016/j.erss.2017.10.005

Kropp, C. (2018). Controversies around energy landscapes in third modernity. *Landscape Research, 43*(4), 562–573. https://doi.org/10.1080/01426397.2017.1287890

Labussière, O., & Nadaï, A. (Eds.). (2018). *Energy transitions: A socio-technical inquiry.* Springer.

Ladenburg, J., & Dubgaard, A. (2007). Willingness to pay for reduced visual disamenities from offshore wind farms in Denmark. *Energy Policy, 35*(8), 4059–4071. https://doi.org/10.1016/j.enpol.2007.01.023

Lennon, B., Dunphy, N., Gaffney, C., Revez, A., Mullally, G., & O'Connor, P. (2019). Citizen or consumer? Reconsidering energy citizenship. *Journal of Environmental Policy & Planning*, 1–14. https://doi.org/10.1080/1523908X.2019.1680277

Lothian, A. (2008). Scenic perceptions of the visual effects of wind farms on South Australian landscapes. *Geographical Research, 46*(2), 196–207. https://doi.org/10.1111/j.1745-5871.2008.00510.x

Marková, I. (2003). *Dialogicality and social representations: The dynamics of mind*. Cambridge University Press.

McLachlan, C. (2009). 'You don't do a chemistry experiment in your best china': Symbolic interpretations of place and technology in a wave energy case. *Energy Policy, 37*(12), 5342–5350. https://doi.org/10.1016/j.enpol.2009.07.057

Moloney, G., Gamble, M., Hayman, J., & Smith, G. (2015). Without anchor: Themata and blood donation. *Papers on Social Representations, 24*(2), 2–1. http://psr.iscte-iul.pt/index.php/PSR/article/download/116/80

Moscovici, S. (1961/72). *La psychanalyse, son image et son public*. Presses universitaires de France.

Naumann, M., & Rudolph, D. (2020). Conceptualizing rural energy transitions: Energizing rural studies, ruralizing energy research. *Journal of Rural Studies, 73*, 97–104. https://doi.org/10.1016/j.jrurstud.2019.12.011

Phadke, R. (2011). Resisting and reconciling big wind: Middle landscape politics in the New American West. *Antipode, 43*(3), 754–776. https://doi.org/10.1111/j.1467-8330.2011.00881.x

Phillips, M., Fish, R., & Agg, J. (2001). Putting together ruralities: Towards a symbolic analysis of rurality in the British mass media. *Journal of Rural Studies, 17*(1), 1–27. https://doi.org/10.1016/S0743-0167(00)00015-2

Seamon, D. (2014). Place attachment and phenomenology: The synergistic dynamism of place. In L. Manzo & P. Devine-Wright (Eds.), *Place attachment: Advances in theory, method and application* (pp. 11–22). Routledge.

Sevenant, M., & Antrop, M. (2010). The use of latent classes to identify individual differences in the importance of landscape dimensions for aesthetic preference. *Land Use Policy, 27*(3), 827–842. https://doi.org/10.1016/j.landusepol.2009.11.002

Sibille, A. D. C. T., Cloquell-Ballester, V. A., Cloquell-Ballester, V. A., & Darton, R. (2009). Development and validation of a multicriteria indicator for the assessment of objective aesthetic impact of wind farms. *Renewable and Sustainable Energy Reviews, 13*(1), 40–66. https://doi.org/10.1016/j.rser.2007.05.002

Silva, L. (2013). Concluding remarks on perceived and lived ruralities and the future of rural Europe. In L. Silva & E. Figueiredo (Eds.), *Shaping rural areas in Europe: Perceptions and outcomes on the present and the future* (pp. 247–254). Springer.

Silva, L., & Prista, M. (2016). Social differentiation in the consumption of a pastoral idyll through tourist accommodation: Two Portuguese cases. *Journal of Rural Studies, 43*, 183–192. https://doi.org/10.1016/j.jrurstud.2015.12.006

Toke, D., Breukers, S., & Wolsink, M. (2008). Wind power deployment outcomes: How can we account for the differences? *Renewable and Sustainable Energy Reviews, 12*(4), 1129–1147. https://doi.org/10.1016/j.rser.2006.10.021

Tonts, M., & Greive, S. (2002). Commodification and creative destruction in the Australian rural landscape: The case of Bridgetown, Western Australia. *Australian Geographical Studies, 40*(1), 58–70. https://doi.org/10.1111/1467-8470.00161

Tveit, M., Ode, Å., & Fry, G. (2006). Key concepts in a framework for analysing visual landscape character. *Landscape Research, 31*(3), 229–255. https://doi.org/10.1080/01426390600783269

Wagner, W. (1998). Social representations and beyond: Brute facts, symbolic coping and domesticated worlds. *Culture & Psychology, 4*(3), 297–329. https://doi.org/10.1177/1354067X9800400302

Wagner, W., & Kronberger, N. (2001). Killer tomatoes! Collective symbolic coping with biotechnology. In K. Deaux & G. Philogene (Eds.), *Representations of the social – Bridging theoretical traditions* (pp. 147–164). Blackwell.

Wallwork, J., & Dixon, J. A. (2004). Foxes, green fields and Britishness: On the rhetorical construction of place and national identity. *British Journal of Social Psychology, 43*(1), 21–39. https://doi.org/10.1348/014466604322915962

Wolsink, M. (2000). Wind power and the NIMBY-myth: Institutional capacity and the limited significance of public support. *Renewable Energy, 21*(1), 49–64. https://doi.org/10.1016/S0960-1481(99)00130-5

Wolsink, M. (2018). Co-production in distributed generation: Renewable energy and creating space for fitting infrastructure within landscapes. *Landscape Research, 43*(4), 542–561. https://doi.org/10.1080/01426397.2017.1358360

Woods, M. (1998). Researching rural conflicts: Hunting, local politics and actor-networks. *Journal of Rural Studies, 14*(3), 321–340. https://doi.org/10.1016/S0743-0167(97)00038-7

Woods, M. (2005). *Rural*. Routledge.

Wüstenhagen, R., Wolsink, M., & Bürer, M. J. (2007). Social acceptance of renewable energy innovation: An introduction to the concept. *Energy Policy, 35*(5), 2683–2691. https://doi.org/10.1016/j.enpol.2006.12.001

PART III

Repoliticisations—Recognising and Articulating Power Relations

CHAPTER 4

Getting Used to It, But …? Rethinking the Elusive U-Curve of Acceptance and Post-Construction Assumptions

David Rudolph and Laura Tolnov Clausen

Introduction

> Fortunately, solutions can be found. Other studies show that fear is biggest when a new [renewable energy] plant is in the planning phase. Once the wind turbine has been established, many of the concerns and opposition disappear. (Dansk Energi, 2019, p. 14)

> [E]ven though we say, well, we have grown accustomed to it, yes, yes, but it is still the first thing I see when I open the door. (Interviewee referring to wind turbines, Denmark, 2018)

D. Rudolph (✉)
Department of Wind Energy, Technical University of Denmark, Roskilde, Denmark
e-mail: dpru@dtu.dk

L. T. Clausen
University of Agder, Kristiansand, Norway
e-mail: laura.t.clausen@uia.no

S. Batel, D. Rudolph (eds.), *A critical approach to the social acceptance of renewable energy infrastructures*,
https://doi.org/10.1007/978-3-030-73699-6_4

A common understanding of authorities and developers to public resistance against renewable energy technologies (RET), in particular wind farms, is that local citizens will get used to them over time. Such an understanding entails a temporal dimension and a rationale of time being able to heal all sores and relativises fears and alleviates concerns once a project has been deployed. This rationale is also reflected in the first quote from a Danish business and interest organisation for energy companies. Whilst such an understanding acknowledges public resistance as a real challenge to the deployment of RET solutions, it also includes a form of consensus between planning, business and research in relation to the underlying understanding of time as an expression of conflict resolution. This is also reflected in early social acceptance research hinting at a necessarily dynamic interplay between development stages and attitudes. In particular, the so-called U-shaped curve (Devine-Wright, 2005; Krohn & Damborg, 1999; Wilson & Dyke, 2016; Wolsink, 2007) represents a popular understanding that people's attitudes towards RET are dynamic and change with time (Fig. 4.1). The universal principle has been that, after an often contested consenting process, people's attitudes towards a project become

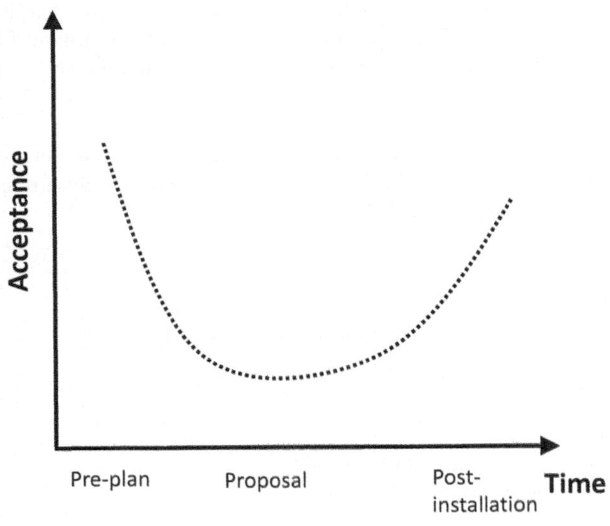

Fig. 4.1 Sketch of U-curve of wind farm attitudes, based on Wolsink (2007) and Ellis and Ferraro (2016)

more favourable once it has been established, thereby presuming some sort of adaptation or familiarisation with a project in particular and change in general. Initial support drops during the planning stages when fears over negative impacts are aroused and acceptance increases again "if it becomes apparent that the change has less impact than expected" (Wheeler, 2016, p. 127). However, the interchangeable use of acceptance and support tends to blur not only differences between passive reaction and active engagement, but also more nuanced responses to RET (e.g. Batel et al., 2013; Bauwens & Devine-Wright, 2018). These semantic intricacies also resonate with people's utterances as reflected in the second quote above echoing the common phrase of 'getting used to it'.

While the dynamic interpretation mirrored by the U-curve correctly moves away from more static understandings of local acceptance that tend to capture a snapshot in time, subsequent research has mainly focused on the planning, development and consenting stages as the problematic phases in which acceptance issues are to be addressed and influenced. Although this has greatly contributed to the generation of valuable insights into the underlying entanglements of opposition and acceptability, it has, in turn, not only resulted in rather instrumental practices to nurture greater acceptance, but also produced resurgent and unchallenged assumptions of what is happening after the deployment of a project. Hence, just as the term 'acceptance' cannot be equated with support and obscures more nuanced responses (Batel et al., 2013), familiarisation and adaptation cannot simply be taken for granted and may instead factor out further emotive and practical reactions between denial, apathy, lack of interest, indifference, silence, tolerance, co-existence, happiness as well as other expressions of coping. In a nutshell, adaptation or familiarisation should not be confused with (greater or regained) acceptance. The installation of an infrastructure project does not constitute an all-encompassing end-point; instead it depicts a new beginning for local residents that popular, yet simplistic presumptions inherent in the U-shaped curve fail to acknowledge adequately. Even more so, presuming the ubiquity of adaptation and familiarisation involves the danger of obscuring anticipatory practices that bring together the ends (i.e. a desired and acceptable future object or condition) and means (i.e. planning and development of the condition here and now).

With the following explorative consideration we do not intend to assess the myriad of post-construction reactions, but attempt to unravel the unarticulated temporalities of social acceptance by fathoming what may be hidden in the protective and relativising phrase of 'getting used to it, *but* ...'. In order to evade potential pitfalls of short-sighted assumptions of adaptation and familiarisation, we juxtapose two notions of how the future is conceived and acted on in the presence. In doing so, we argue for less instrumental visions of locally desired and beneficial renewable energy projects based on prefigurative politics.

In the following, we will first shed light on the inherent temporalities of social acceptance research, before outlining some fallacies of persistent post-construction assumptions. We then hint at what may be hidden behind a taken-for-granted familiarisation with or adaptation to RET projects and finally sketch out some thoughts of how ends and means of planning can be brought together more fruitfully.

Temporality within Social Acceptance Research

Time is an essential condition in research dealing with the social acceptance of RET. Despite the immense literature replete with implicit references to temporal conditions, the significance of time in the formation and governance of local responses to renewables has, however, remained largely unexamined. Yet, the significance of exploring the meaning of time in responses to RET (Batel, 2018), both as a relational factor (Labussière & Nadaï, 2018) and as diachronic processes in energy transition (Gismondi, 2018), has recently been highlighted. In this section we briefly outline domains in which temporality becomes relevant in social acceptance research.

It is not surprising that one of the main interests in acceptance research has been on the consenting stages of RET planning, as this is when most often looming controversies and conflicts erupt. Planning is a temporal process and, in turn, temporality is a fundamental feature of planning. It does not merely refer to the process of devising procedures for how to achieve a desired goal, but also to the legal context for permitting infrastructure developments. In the latter sense, planning includes a process of assessing (or even predicting) outcomes and subsequently arranging mitigatory measures for potential environmental and social impacts that may occur during and after the construction of a RET project, as exemplified in environmental impact assessments (EIA). Various procedural features within planning and development stages, such as recognition and engagement of affected communities, have been identified as constitutive elements of community responses to proposed developments (e.g. Haggett,

2014; Ottinger et al., 2014; Rydin et al., 2018). Likewise, distributional features of acceptability, such as the anticipated distribution of impacts and benefits during and after construction, bear a distinct temporal characteristic, in particular when juxtaposing immediate local impacts of RET projects with moral obligations in terms of distant, intangible and long-term benefits related to climate change mitigation. Thus, the temporality of distributional justice also relates to specific questions of when potential local benefits fostered through infrastructure projects are delivered or come into effect. Moreover, RET projects as such, and the landscapes they (co)produce, encompass a distinct temporal dimension. Sherren et al. (2016) have explored how residents have come to appreciate benefits of long-term landscape change induced by hydropower over time. Windemer (2019) has raised the important question to what extent the understanding and impression of wind farms as temporary infrastructures with time-limited consents, expiring permits and open end-of-life considerations shape local responses and expectations. Similarly, Pasqualetti and Stremke (2018) have drawn attention to the consequences for the perception and permanence of energy landscapes, whose temporal characteristics may be either relatively dynamic and transient due to a short life cycle of technologies and the reversibility of interventions or more permanent and irreversible.

As a particular dimension of such perceptions, local resistance to RET projects or spatial changes can be construed as attempts to avert or mitigate the occurrence of a threat or to retain continuity. Thus, opposition holds assumptions about the future, future events or future conditions, as fundamentally reflected in the notion of place-protective action (Devine-Wright, 2009). However, not only rationales of opposition, but the entire concept of social acceptance entails a temporal logic that is ultimately directed to the things to come. Acceptance, accepting communities or the acceptability of energy infrastructures and energy landscapes are constituted through anticipatory practices of the future.

In particular, local acceptance in terms of non-oppositional and affirmative relationships between a community and a RET cannot simply be taken for granted, but has to be acquired and nurtured, and is thus fundamentally directed to the future. Acceptance is something that is aimed at being achieved in the future by seeking to set the 'right' conditions and making the 'right' decisions in the presence. These acceptance-shaping conditions are most profoundly reflected in managerial practices of gaining trust (e.g. Fast & Mabee, 2015; Gross, 2007), the idea of sharing benefits (e.g. Cowell et al., 2011), and participation in planning and decision-making (e.g. Aitken et al., 2016; Natarajan et al., 2018) that are all aimed at creating acceptable outcomes. Striving for acceptance can therefore be regarded as a form of anticipatory action, as a way of acting not only in advance of, but also on the basis of the future. This future is, however, largely unknown, indeterminate and exceeds present knowledge, which renders anticipatory action rather directed towards an expected, predicted or desired future condition. Thus, a, in whatever way, envisioned "future becomes cause and justification for some form of action in the here and now" (Anderson, 2010, p. 778). In other words, the anticipated ends become entangled with their means through justifications, that is, the particular means are justified with the ends they are aimed at. Anderson (2010) describes three modes of practices (calculation, imagination and performance) through which future conditions are made graspable while remaining absent and three logics (preemption, precaution and preparedness) through which futures are acted on. Preemption, precaution and preparedness are all means of guiding action towards a future that has been problematised and made present in a certain way (Anderson, 2010), most commonly as disruptive to a desired end, for example, local opposition to a proposed wind farm project. In particular, the notions of preemption and preparedness have been inherent in practices of dealing with issues related to local acceptance. Preemption entails logics of intervention on the basis of indeterminate potentiality, rendering preemptive action as formative in terms of acting over threats that have not yet emerged and unleashing transformative events in the presence in order to prevent the occurrence of a future condition by halting or redirecting a process or initiating a new process. The logic of preparedness, on the other hand, is to prepare for and deal with the aftermath of future events. Preparedness is aimed at averting or alleviating the consequences of a precipitating event (Anderson, 2010, pp. 790–791). Such preemptive and preparatory practices as anticipatory action intended to obtain acceptability of or prevent resistance to a proposed project are mirrored in the provision of participatory opportunities, benefits and proactive or compensatory measures and the mitigation of anticipated impacts determined during the EIA process. In other words, imaginations about the future, regardless of how they are rendered knowable, are constitutive of how they are acted upon and dealt with in the presence. This also pertains to suppositions originating from the so-called U-curve of acceptance, assuming that local acceptance increases after the completion of the project.

The Elusive U-Curve

As mentioned at the beginning, early studies on the social acceptance of renewable energy hinted at a regain of temporarily declined local acceptance after the closure of often contested planning processes and the completion of the development, a change which has been illustrated with the so-called U-shaped curve (e.g. Wolsink, 2007; Fig. 4.1). Surprisingly, despite an ever-increasing research interest in and more nuanced understanding of underlying drivers of acceptance, post-construction acceptance has rarely been addressed nor questioned. However, the perpetuation of U-curve assumptions has arguably become problematic in several regards.

Methodologically, there is a lack of longitudinal studies 'monitoring' the 'acceptance' of operational projects. While numerous studies have referred to the U-shaped curve (see above), only a few studies (e.g. Eltham et al., 2008; Hoen et al., 2019; Mills et al., 2019; Motosu & Maruyama, 2016; Wheeler, 2016; Wilson & Dyke, 2016) have explicitly set out to investigate the post-implementation phase and inquired into how attitudes towards actually existing projects, mainly wind farms, have turned out or changed over time. The few studies looking at attitudes towards operational RET projects are almost exclusively based on quantitative methods (except Wilson & Dyke, 2016; Wheeler, 2016), like surveys, only allowing for limited insights into people's interactions and relationships with existing energy projects and what different expressions of post-implementation acceptance may mean, imply or exclude. Nevertheless, the existing literature generally suggests that familiarity can lead to contentment. Eltham et al. (2008) argued that unfamiliarity with wind turbines can correlate with negative and outdated preconceptions of the technology, while opposition often decreases for those directly affected by existing wind power facilities. Ellis & Ferraro (2016, p. 27) conclude from the rare and often case-specific studies that familiarity may indeed lead to recovered acceptance, but rebut the simplistic idea of communities becoming highly positive when living with wind farms. Instead it is emphasised that a myriad of lived experiences as well as wider social and political forces affect the dynamics of local acceptance, thus contesting an implicit increase of acceptance after construction.

A particular fallacy inherent in unreserved post-construction assumptions relates to the misrecognition of an advanced socio-political context. Since early studies on social acceptance emerged, the RET sector has grown and matured, and especially wind farms are no longer novel developments in many areas. Instead, they have become critical infrastructures, as they have increased in size and number. The emergence of offshore wind farms may have also affected the way onshore wind farms are viewed. The deployment practices have become more commercialised and support systems for renewables have become more competitive, with the consequence of differentiated ownership structures and limited opportunities of collective participation in wind energy developments (e.g. Kirkegaard et al., 2021). Changed socio-political conditions question whether familiarisation with a RET, now as before, leads to contentment over time. Simplistic assumptions of post-construction adaptation tend to be deceptive in procedural aspects, too. In practice, public engagement is front-loaded and takes place during the consenting process, while communication channels facilitating a continued engagement or general assessments of specific concerns after the construction of a project are rarely in place. Disregarding procedural limitations may hone the cessation of institutionalised spaces for argumentation, consensus seeking and precautionary practices down to regained acceptance, or may even confuse the absence of antagonistic efforts with silent acceptance. However, most notably, existing research has shown that conflicting issues and contestations rarely relate to the mere presence of an RET project per se, but are manifold, entangled and involve procedural, distributional and spatial issues, which are not easily reducible to adaptation and familiarity (e.g. Mills et al., 2019).

A preservation of the fallacious assumption of regained acceptance ultimately entails a pitfall that is seemingly characterised by a self-fulfilling prophecy. Premature and unquestioned presumptions of post-implementation acceptance in terms of adaptation and familiarisation over time, that is, of people getting used to change, may easily be distorted to legitimise and open the door for further developments in the future. Hence, this may lead to both a normalisation of past procedures that accompany depoliticised acceptance (Dunlap & Sullivan, 2020) and a suppression of social conflict as valuable participation (Cuppen, 2018). In turn, focusing on pre-installation stages merely gives a snapshot in time that is concerned with whether a RET project attains local acceptance or not. Thus, preemptive principles of planning resulted in (mainly instrumental) approaches to accomplish acceptable projects. However, both the anticipation of people 'getting used' to change and the active effectuation of increased local acceptance through engagement and concessions may obfuscate several critical issues, three of which will be briefly outlined in the next section.

What 'Getting Used to it, but ...' May Hide?

> After all, you can't be mad every day, so you are trying to ignore it. But then, it is still the same. I don't think it belongs here. (Interviewee referring to wind turbines, Denmark, 2018)

Does the termination of an often contested planning phase and the completion of a project represent a fixed endpoint after which nothing is happening or known? Given its short-term managerial, regulatory and preemptive premises, the installation of a RET project depicts a processual endpoint for planning, but for the people in the vicinity of the installation the completion also means a new beginning. While it usually means the end of the period where they can exert any influence, it also connotes the dawn of a new period of living with the infrastructure project. As the quote indicates, 'sharing' a landscape with certain infrastructures can involve lengthy habituation processes which are more characterised by necessity than naturalness. Hence, whereas the presumption of wind farms becoming more acceptable or people getting used to change over time have rarely been questioned, the studies that do so can be characterised by more fundamentally doubting definite links between time and acceptance. For instance Devine-Wright (2005, p. 131) has argued that "it is unlikely that there will be a simple, linear relationship between experience and perception because of the numerous other influences that shape people's judgements and opinions". It is exactly these other influences which we claim to become blurred through short-sighted or hopeful anticipations of familiarisation and adaptation, not only during the consenting process, but also after the installation. Hence, we suggest that there are at least three entangled and formative conditions that become concealed behind protective and stinted 'getting used to' rhetoric and that remain unarticulated when simply referring to familiarisation and adaptation: ***planning procedures***, ***democratic shortcomings*** and ***alienation***.

First, 'getting used to it' or adaptation necessitate some sort of temporal process. The need to get used to something hints at the incongruence between a predicted and an actual outcome, which in turn may point to inadequacies of the *planning procedures* to deal with certain issues. Logics and tools of planning are based on predictions, calculation, quantifications and prior experiences, whereas social impacts and repercussions are more difficult to assess (Larsen et al., 2018) leaving certain voices inadequately or unequally addressed. There is a limit of predictive practices within the EIA process and the extent to which experiences and perceptions of impacts can be forecasted. Although the approach to planning is generally consensus-oriented, some issues, impacts and risks can therefore only be handled through preemptive compensatory and concessionary measures. Moreover, altered attitudes towards proposed and operational RET may result "from an increased acceptance over time as perceived risks do not materialize (and as resignation sets in)" (Wheeler, 2016, p. 127). Of course, the latter explanation is only ostensibly conducive of increased post-construction acceptance. This obscure causality between resignation and acceptance rather bespeaks the ambivalence and vagueness of the term acceptance and hints at more nuanced lived experiences. However, procedures, power relations and features of the planning phase can drag on to post-construction stages (Mills et al., 2019) and hide behind adaptation as resignation. For example, Pohl et al. (2012) and Rudolph et al. (2017) have shown that the experienced deficiencies during the planning process can manifest in and accentuate the subsequent annoyance by impacts of RET.

Second, temporality becomes also relevant with regard to ***democratic deficiencies***. The argument of implied or preemptive acceptance of RET projects is often justified with a preceding process of public involvement where local citizens are given an opportunity to have a say prior to the installation. However, it can be argued that existing procedures of 'invited participation' (Gaventa, 2006) concede citizens a constrained scope for participation (Cuppen, 2018). Not only does the framing of the debate rarely facilitate a deliberative and reflexive 'free space' (Evans & Boyte, 1992) in which alternative reflections, critiques and hopes are evoked in a social dialogue. The legitimising practice of inviting local residents only after specific project proposals have been prepared also constrains the deliberative capacities of communities. Similarly, Janhunen et al. (2018) demonstrate that the need for participation and communication does not expire after the planning process has ended. Even more so, as mentioned, the post-deployment phase is characterised by a lack of a common public space where experiences of living with the infrastructure project can be collectively shared or transformed. There are usually no meaningful channels for long-term engagement, through which local residents can further exercise influence or communicate concerns. Hence, lacking procedures for engagement and limited ways of responding to actually occurring issues after the completion of a project may result in resignation and apathy of people or necessitate rather drastic measures, such as opting to buy properties and displacing residents (Clausen & Rudolph, 2019). Seen from this perspective the notion of adaptation may also hide the effects of powerlessness in the process of 'getting used to it'.

Third, procedural and distributional deficiencies may manifest in processes of ***alienation***. In particular, alienation can be seen as a product of capitalist enclosure of land and advanced commodification of everything (see Chatterton & Pusey, 2020; Harvey, 2018), including RET infrastructures, that are increasingly impelled by cost reductions, profit maximising and capacity expansions while marginalising local considerations. In practice, this situation is amplified by the condition that ownership of both the land and ever-larger RET infrastructures do rarely belong (any longer) to those being affected the most (Clausen & Rudolph, 2019; Kirkegaard et al., 2021). Such formative processes of local residents have also been referred to as accumulation-by-alienation (Dunlap & Sullivan, 2020), implying that the advancement of large-scale RET facilities has increasingly detached local people from possibilities of material and affective ownership. This form of alienation has prompted governments to provide mainly instrumental measures aimed at reattaching people to RET facilities through proactive benefit-sharing or financial participation. However, the more instrumental the planning, the more alienating it tends to be for those affected. Thus, adaptation may also hide a state of indifference due to alienating effects of power relations rebounding from the planning process and distributional injustices. In contrast to what the U-curve may suggest, post-construction acceptance cannot simply be taken for granted or hoped for, just because previous contestations abate. The process of getting used to change may rather consist of practices that help affected people to cope with the changes themselves, but also with the preceding terms and conditions that effectuated the change and potentially alienating effects in order to preserve certain people-place relationships.

Mindful of the limitations of oversimplification of these admittedly fragmentary considerations, we beg to nuance: The processes of 'getting used to it' (let alone regained acceptance), is likely to entail elements of coping (that may still spawn adaptation in the long run), rather than being an auspicious matter of adaptation and familiarisation (see also Devine-Wright, 2009). However, a so far largely missing critical perspective on post-construction experiences can shed light on these nuances.

From Preemptive Acceptance to Prefiguration

How could then a desired future outcome be achieved without (completely) lapsing back into preemptive and precautionary principles or invoking instrumental and techno-managerial practices of deploying RET facilities that primarily employ rationales of creating accepting citizens, acceptable locations and technologies from a distance? How can the passivity of 'getting used to it', democratic shortcomings and rising alienation from the outcome be warded off in practice? Inspiration for the formation of a locally desired and beneficial renewable energy future can be drawn from the concepts of radical planning and prefigurative politics. Both concepts put affected citizens at the beginning and the heart of the development of RET projects.

Radical planning challenges the monopoly of the state, its institutions and corporate actors over regulatory governance and thus the legitimate production and conceptualisation of space for development (Miraftab, 2009). The rationale of radical planning is to follow one's own creativity, needs and desires as a starting point instead of trying to demand the state to act in a certain way or to rectify inequalities. Since organisational functions cannot be fully relinquished, this does not mean to obviate the necessity of the state, but to bypass instrumental, managerial and standardised approaches of the *status quo* as a means to ensure desired outcomes or forestall an undesired future, thus rendering dominant procedures and structures redundant by showing and following another way (Hillier, 2017). As suggested by Miraftab (2009), it involves purposeful actions that aim to destabilise the *status quo* through consciousness of the past and imagination of an alternative future, such as insurgent spaces to overcome democratic shortcomings. A distinctive element of radical planning is the concept of prefiguration or prefigurative politics originating from anarchist traditions (Hillier, 2017; Ince, 2012; Purcell, 2017). Anarchist prefigurative politics resides in the contestations, struggles and counter-institutional practices of everyday life and "emphasizes the ongoing development of the practices and relations of envisioned future worlds in the present" (Ince, 2014, p. 278). The core of prefigurative politics is a generative temporal framing that unifies ends and means, wherein any future state is seen as indeterminate and the product of practices, affordances, and contingencies preceding it (Gordon, 2018). In contrast to a pragmatist unification of ends and means based on mutual definition and reciprocity (Labussière & Nadaï, 2018), prefiguration drops the distinction and rather collapses the ends (future) into the means (present). Thus, prefiguration "subverts the politics of waiting by embracing the immediate possibilities of the *here* and *now*" (Springer, 2016, p. 164). It therefore disbands the temporal distinction between the practices in the present and distant goals in the future, "bridging past, present and future within an unfolding plan (Gordon, 2018, p. 526). In doing so, prefigurative politics invert classical "strategic politics as ordering tactics towards a particular goal" (Swain, 2019, p. 51), as reflected in the EIA procedures or measures aimed at future acceptance among residents, and share the understanding that a particular end is not predetermined but instead mirrored and embodied in its means in a direct and immediate way. Yet, it is not only about the practices themselves that prefigure, but also about the organisation and processes behind the acts that foretaste the envisioned future (Ince, 2012; Swain, 2019). Prefiguration incorporates non-alienating modes of interaction based on the fruition of shared values in the present. Ends-effacing prefiguration is capable of bringing together various people with diverse interests to continuously enact objectives "in an enduring process of change" (Labussière & Nadaï, 2018, p. 312). Reconceiving acceptance and adaptation in prefigurative terms, means envisioning a desired future not only as something to fight for in the present to be obtained in the future, but as something that is already here, something that should be embraced in the present and something that is negotiated by all affected actors.

In summary, preemptive or precautionary vis-à-vis prefigurative principles entail two fundamentally different logics for bringing together the ends and means of RET development. While preemption, precaution and preparedness tend to employ reactive and preventive logics, prefiguration evokes a more active and assertive logic of practice. Within precautionary principles of planning, the anticipated ends are reflected in the means through justifications to prevent or prepare for certain ends, such as mitigatory measures of potentially adverse impacts and managerial means to achieve widespread acceptance or avert resistance. In contrast, prefigurative logics envision desired and valued ends in the here and now and unify the outcome with the means to achieve it, as a concrete utopia (see also Brand & Wissen, 2018, p. 84). Even if both principles can be regarded as anticipatory practices, as acting on the future in the presence by bringing together the ends and means, only prefigurative rationales appear to be truly generative towards a desired future and capable of bringing about locally desired, beneficial and non-alienating outcomes from the very beginning. This is because the essential difference is in the nature of the objectives and how the future is conceived. Preemption and precaution emphasise the purposeful implementation of a goal (i.e. the deployment of a wind farm) and conceive the future as contingent and uncertain, whereas prefiguration embodies an envisioned future and apprehends the goal as a means to achieve further desirable goals or to address certain needs.

Incorporating notions from prefigurative politics in RET development and planning opens up new logics of RET development with consequences for social acceptance research and practice. Instead of assuming, hoping for, working towards or instrumentalising post-construction acceptance as upheld by rationales inherent the U-curve, the idea would be to create a desirable and non-alienating project in a more inclusive democratic process in the here and now by starting from local requirements, and not simply together with, but originating from communities of interest and place. This would imply a utilisation of RET technologies scaled to local needs, values and collective benefit-sharing and fundamentally mean to "work outwards from the place and not inwards from the system" (Devine-Wright, 2019, p. 894). In theory, this would then also invert the object of acceptance, from acceptance among the local residents, that is, accepting people, to enacting locally meaningful projects by working towards a socio-spatial fit. In practice, this could entail citizen assemblages or juries and invited spaces for developers.

Conclusions

A fundamental challenge of low-carbon energy transition is grounded in the recurring finding of previous research that controversies over the deployment of RET facilities often comprise issues that reach beyond the technology- and energy-related topics per se and point to the wider social context. Thus, acceptance issues cannot simply be reduced to either blemishes of the object at stake or the intransigence of affected people. Given these circumstances, this chapter aimed to shed critical light on the obscurities of a fallacious perseverance of presumptions directed to post-construction acceptance and to sketch out theoretical underpinnings for a pathway to more inclusive, fairer and locally embedded RET developments. The conclusions of the previous considerations are two-fold. First, acceptance after construction cannot simply be taken for granted just because previously heated situations have seemingly abated after the completion, people have noticeably ceased to object or claim to have grown used to changes. This may well be the case, but ambivalent findings of limited research rather indicate that these circumstances cannot be equated with rising acceptance, as oversimplified with the U-shaped curve. Instead there is an urgent need for more qualitative research looking into the temporalities of local acceptance, how people's interactions and relationships with operational RET projects and altered places actually play out and what this, in turn, means for post-carbon societies, the planning process and development practices. This includes research focusing on behavioural, psychological and discursive elements to assess how people cope with operational projects, but also research focusing on the temporalities of the formation of acceptance drawing on historical materialism, actor constellations and power relations. The latter perspective can help examine how post-construction conditions and low-carbon energy geographies have been 'produced' and influence the interplay of people, technology and place. At the very least, this would provide new insights into what 'getting used to' may mean and imply. Second, in practical terms, local controversies over RET should not solely be regarded as an issue of acceptance but can provide an opportunity for challenging established principles. The planning, development and deployment of RET facilities do not only have to take account of the (local) social context, but would have to start from the social context in order to realise locally meaningful, beneficial and transformative projects. As suggested, such a radical shift could draw on ideals and logics of prefigurative 'radical' planning to help shape a positive future in the here and now instead of precautionary, managerial and instrumental practices to effectuate acceptable outcomes. Such an approach that ventures beyond constraints of contingent futures requires imagining together, planning together and acting together in ways that prefigure a post-carbon world we would like to inhabit. The impetus for RET developments would need to emerge from within communities reflecting local requirements and values rather than being imposed from the outside, implemented through standardised procedures and primarily driven by (inter)national priorities. While this rationale has been a central principle in the evolution of various community energy initiatives (Devine-Wright, 2019), its adequacy for large-scale developments is more challenging and remains to be explored. A fundamental challenge arises because envisioning such an alternative development practice for RET projects at large does not only imply an emancipation of communities from their roles as legitimising bystanders. But it has inevitably to "move beyond technocratic elitism" (Springer, 2016, p. 18) and overturn hegemonic ways of planning, financing and ownership that prioritise growth rather than social transformation. However, our own preferences and willingness will probably need to change too, and they are likely to do so only as a result of accompanying others in new kinds of shared ideas. Our explorative thoughts should be viewed as both an entry point for further critical engagements with temporal underpinnings of 'acceptability' and stimuli for activist research. Since means embody desired ends, it is time to embrace the desired 'post-construction' future here and now.

References

Aitken, M., Haggett, C., & Rudolph, D. (2016). Practices and rationales of community engagement with wind farms: Awareness raising, consultation, empowerment. *Planning Theory & Practice, 17*(4), 557–576. https:f4649357.2016.1218919

Anderson, B. (2010). Preemption, precaution, preparedness: Anticipatory action and future geographies. *Progress in Human Geography, 34*(6), 777–798. https://doi.org/10.1177/2F0309132510362600

Batel, S., Devine-Wright, P., & Tangeland, T. (2013). The social acceptance of low-carbon energy and associated infrastructures: A critical discussion. *Energy Policy, 58*, 1–5. https://doi.org/10.1016/j.enpol.2013.03.018

Batel, S. (2018). A critical discussion of research on the social acceptance of renewable energy generation and associated infrastructures and an agenda for the future. *Journal of Environmental Policy & Planning, 20*(3), 356–369. https://doi.org/10.1080/1523908X.2017.1417120

Bauwens, T., & Devine-Wright, P. (2018). Positive energies? An empirical study of community energy participation and attitudes to renewable energy. *Energy Policy, 118*, 612–625. https://doi.org/10.1016/j.enpol.2018.03.062

Brand, U., & Wissen, M. (2018). *The limits to capitalist nature. Theorizing and overcoming the Imperial mode of living*. Rowman & Littlefield.

Chatterton, P., & Pusey, A. (2020). Beyond capitalist enclosure, commodification and alienation: Postcapitalist praxis as commons, social production and useful doing. *Progress in Human Geography, 44*(1), 27–48. https://doi.org/10.1177/2F0309132518821173

Clausen, L. T., & Rudolph, D. (2019). (Dis)Embedding the wind—On people-climate reconciliation in Danish wind power planning. *The Journal of Transdisciplinary Environmental Studies, 17*(1), 5–21.

Cowell, R., Bristow, G., & Munday, M. (2011). Acceptance, acceptability and environmental justice—The role of community benefits in wind farm development. *Journal of Environmental Planning and Management, 54*(4), 539–557. https://doi.org/10.1080/09640568.2010.521047

Cuppen, E. (2018). The value of social conflicts. Critiquing invited participation in energy projects. *Energy Research & Social Science, 38*, 28–32. https://doi.org/10.1016/j.erss.2018.01.016

Dansk Energi. (2019). Renewable Energy Outlook 2019. *Renewable Energy Perspectives to 2035.* February 2019. https://www.danskenergi.dk/sites/danskenergi.dk/files/media/dokumenter/2019-06/Renewable_Energy_Outlook_2019.pdf

Devine-Wright, P. (2005). Beyond NIMBYism: Towards an integrated framework for understanding public perceptions of wind energy. *Wind Energy, 8*, 125–139. https://doi.org/10.1002/we.124

Devine-Wright, P. (2009). Rethinking NIMBYism: The role of place attachment and place identity in explaining place-protective action. *Journal of Community & Applied Social Psychology, 19*(6), 426–441. https://doi.org/10.1002/casp.1004

Devine-Wright, P. (2019). Community versus local energy in a context of climate emergency. *Nature Energy, 4*, 894–896. https://doi.org/10.1038/s41560-019-0459-2

Dunlap, A., & Sullivan, S. (2020). A faultline in neoliberal environmental governance scholarship? Or, why accumulation-by-alienation matters. *Environment and Planning E. Nature and Space, 3*(2), 552–579. https://doi.org/10.1177/2F2514848619874691

Ellis, G., & Ferraro, G. (2016). *The social acceptance of wind energy. Where we stand and the path ahead. JCR science for policy report. European Commission.* Publication office of the European Union.

Eltham, D. C., Harrison, G. P., & Allen, S. J. (2008). Change in public attitudes towards a Cornish wind farm: Implications for planning. *Energy Policy, 36*(1), 23–33. https://doi.org/10.1016/j.enpol.2007.09.010

Evans, S. M., & Boyte, H. C. (1992). *Free spaces: The sources of democratic change in America.* University of Chicago Press.

Fast, S., & Mabee, W. (2015). Place-making and trust-building: The influence of policy on host community responses to wind farms. *Energy Policy, 81*, 27–37. https://doi.org/10.1016/j.enpol.2015.02.008

Gaventa, J. (2006). Finding the spaces for change: A power analysis. *IDS Bulletin, 37*(6), 23–33. https://doi.org/10.1111/j.1759-5436.2006.tb00320.x

Gismondi, M. (2018). Historizing transitions: The value of historical theory to energy transition research. *Energy Research & Social Science, 38*, 193–198. https://doi.org/10.1016/j.erss.2018.02.008

Gordon, U. (2018). Prefigurative politics between ethical practice and absent promise. *Political Studies, 66*(2), 521–537. https://doi.org/10.1177/2F0032321717722363

Gross, C. (2007). Community perspectives of wind energy in Australia: The application of a justice and community fairness framework to increase social acceptance. *Energy Policy, 35*(5), 2727–2736. https://doi.org/10.1016/j.enpol.2006.12.013

Haggett, C. (2014). Planning and persuasion: Public engagement in renewable energy decision-making. In P. Devine-Wright (Ed.), *Renewable energy and the public. From NIMBY to participation* (pp. 15–27). Earthscan.

Harvey, D. (2018). Universal alienation. *Journal of Cultural Research, 22*(2), 137–150. https://doi.org/10.1080/14797585.2018.1461350

Hillier, J. (2017). On planning for not having a plan? *Planning Theory & Practice, 18*(4), 668–675. https://doi.org/10.1080/14649357.2017.1369231

Hoen, B., Firestone, J., Rand, J., Elliot, D., Hübner, G., Pohl, J., Wiser, R., Lantz, R., Ryan Haac, T., & Kaliski, K. (2019). Attitudes of U.S. wind turbine Neighbors: Analysis of a Nationwide survey. *Energy Policy, 134*, 110981. https://doi.org/10.1016/j.enpol.2019.110981

Ince, A. (2012). In the Shell of the old: Anarchist geographies of territorialisation. *Antipode, 44*(5), 1645–1666. https://doi.org/10.1111/j.1467-8330.2012.01029.x

Ince, A. (2014). The shape of geography to come. *Dialogues in Human Geography, 4*(3), 276–282. https://doi.org/10.1177/2F2043820614540852

Janhunen, S., Hujala, M., & Pätäri, S. (2018). The acceptability of wind farms: The impact of public participation. *Journal of Environmental Policy & Planning, 20*(2), 214–235. https://doi.org/10.1080/1523908X.2017.1398638

Kirkegaard, J. K., Cronin, T., Nyborg, S., & Karnøe, P. (2021). Paradigm shift in Danish wind power: The (un) sustainable transformation of a sector. *Journal of Environmental Policy & Planning 23*(1), 97–113. https://doi.org/10.1080/1523908X.2020.1799769

Krohn, S., & Damborg, S. (1999). On public attitudes towards wind power. *Renewable Energy, 16*(1), 954–960. https://doi.org/10.1016/S0960-1481(98)00339-5

Labussière, O., & Nadaï, A. (2018). The temporalities of energy transition processes. In O. Labussière & A. Nadaï (Eds.), *Energy transitions: A socio-technical inquiry* (pp. 277–317). Palgrave Macmillan.

Larsen, S. V., Hansen, A. M., & Nielsen, H. (2018). The role of the EIA and weak assessments of social impacts in conflicts over implementation of renewable energy. *Energy Policy, 115*, 43–53. https://doi.org/10.1016/j.enpol.2018.01.002

Mills, S. B., Bessette, D., & Smith, S. (2019). Exploring landowners' post-construction changes in perception of wind energy in Michigan. *Land Use Policy, 82*, 754–762. https://doi.org/10.1016/j.landusepol.2019.01.010

Miraftab, F. (2009). Insurgent planning: Situating radical planning in the global south. *Planning Theory, 8*(1), 32–50. https://doi.org/10.1177/2F1473095208099297

Motosu, M., & Maruyama, Y. (2016). Local acceptance by people with unvoiced opinions living close to a wind farm: A case study from Japan. *Energy Policy, 91*, 362–370. https://doi.org/10.1016/j.enpol.2016.01.018

Natarajan, L., Rydin, Y., & Lock, S. J. (2018). Navigating the participatory processes of renewable energy infrastructure regulation. *Energy Policy, 114*, 201–210. https://doi.org/10.1016/j.enpol.2017.12.006

Ottinger, G., Hargrave, T. J., & Hopson, E. (2014). Procedural justice in wind facility siting: Recommendations for state-led siting processes. *Energy Policy, 65*, 662–669. https://doi.org/10.1016/j.enpol.2013.09.066

Pasqualetti, M., & Stremke, S. (2018). Energy landscapes in a crowded world: A first typology of origins and expressions. *Energy Research & Social Science, 36*, 94–105. https://doi.org/10.1016/j.erss.2017.09.030

Pohl, J., Hübner, G., & Mohs, A. (2012). Acceptance and stress effects of aircraft obstruction lights of wind turbines. *Energy Policy, 50*, 592–600. https://doi.org/10.1016/j.enpol.2012.07.062

Purcell, M. (2017). Our own power to act. *Planning Theory & Practice, 18*(4), 690–694. https://doi.org/10.1080/14649357.2017.1379787

Rudolph, D., Kirkegaard, J., Lyhne, I., Clausen, N. E., & Kørnøv, L. (2017). Spoiled darkness? Sense of place and annoyance over obstruction lights from the world's largest wind turbine test Centre in Denmark. *Energy Research & Social Science, 25*, 80–90. https://doi.org/10.1016/j.erss.2016.12.024

Rydin, Y., Natarajan, L., Lee, M., & Lock, S. (2018). Local voices on renewable energy projects: The performative role of the regulatory process for major offshore infrastructure in England and Wales. *Local Environment, 23*(5), 565–581. https://doi.org/10.1080/13549839.2018.1449821

Sherren, K., Beckley, T. M., Parkins, J. R., Stedman, R. C., Keilty, K., & Morin, I. (2016). Learning (or living) to love the landscapes of hydroelectricity in Canada: Eliciting local perspectives on the Mactaquac dam via headpond tours. *Energy Research & Social Science, 14*, 102–110. https://doi.org/10.1016/j.erss.2016.02.003

Springer, S. (2016). *The anarchist roots of geography*. University of Minnesota Press, Minneapolis.

Swain, D. (2019). Not not but not yet: Present and future in Prefigurative politics. *Political Studies, 67*(1), 47–62. https://doi.org/10.1177/2F0032321717741233

Wheeler, R. (2016). Reconciling windfarms with place identity: Exploring residents' attitudes to existing sites. *Sociologica Ruralis, 57*(1), 110–132. https://doi.org/10.1111/soru.12121

Wilson, G. A., & Dyke, S. L. (2016). Pre- and post-installation community perceptions of wind farm projects: The case of Roskrow Barton (Cornwall, UK). *Land Use Policy, 52*, 287–296. https://doi.org/10.1016/j.landusepol.2015.12.008

Windemer, R. (2019). Considering time in land use planning: An assessment of end-of-life decision making for commercially managed onshore wind schemes. *Land Use Policy, 87*, 104024. https://doi.org/10.1016/j.landusepol.2019.104024

Wolsink, M. (2007). Planning of renewables schemes: Deliberative and fair decision-making on landscape issues instead of reproachful accusations of non-cooperation. *Energy Policy, 35*(5), 2692–2704. https://doi.org/10.1016/j.enpol.2006.12.002

CHAPTER 5

Does Renewable Energy Exist? Fossil Fuel+ Technologies and the Search for Renewable Energy

Alexander Dunlap

Introduction

Fossil fuels are the heart of capitalism, industrialism and state formation, the results of which have been ecologically catastrophic (Malm, 2016).[1] Renewable energy, on the other hand, has emerged as the protagonist of our times, positioned as a solution to our ever-increasing energy consumption and environmental issues. Along with market-based and "natural capital" conservation schemes (see Büscherr & Fletcher, 2020; Dunlap & Sullivan, 2020), renewable energy (RE) systems are promoted as the central mitigating forces against climate change and ecological degradation. Renewable energy, like market-based conservation, is not the solution many of us think it is.

A. Dunlap (✉)
University of Oslo, Oslo, Norway
e-mail: alexander.dunlap@sum.uio.no

S. Batel, D. Rudolph (eds.), *A critical approach to the social acceptance of renewable energy infrastructures*,
https://doi.org/10.1007/978-3-030-73699-6_5

Bernie Sanders' (2019) elaboration of the Green New Deal recognizes the extractive cost of "renewable energy" systems and offers the most comprehensive regulations against the fossil fuel industry (see Dunlap, 2019b). The Green New Deal, however, claims that "it will achieve 100 percent sustainable energy for electricity and transportation by no later than 2030 and will fully decarbonize the economy by 2050" (Sanders, 2019). The bad/good energy dichotomy of fossil fuels versus renewable energy, a holdover from the US Department of Energy and 1970s environmentalism, is misleading, if not entirely false. Little, but carbon accounting calculations, suggest that ecological ecologicla catastrophe will slow from state supported market-based environmental policy schemes. So-called renewables "are becoming fossilized", as Sujatha Raman (2013) points out, reflecting the same geopolitical (ecology) patterns and problems as the hydrocarbon industry. Environmental and climate policy rebrand energy extraction and computational technologies as "environmentally friendly", while widening bike paths and upgrading police stations with solar panels. This infrastructural development is not environmentally friendly or ecologically sustainable, and this chapter argues that, in fact, that there is no such thing as ***renewable energy*** as we know the term. Renewable energy infrastructures themselves comprise of fossil fuels or, at best, are fossil fuel+ technologies.

"Fossil fuel+ technologies" is the more appropriate term for renewable energy. Industrial and utility-scale energy systems represent a structural socio-ecological problem—and (national security) policy issue—that requires greater acknowledgement. This also includes the increasing need for renewable energy's "social acceptance" scholars (see Creamer et al., 2019; Devine-Wright, 2019), as suggested by Susana Batel (2020), to question further the so-called sustainability and energetic renewability of these infrastructural systems. The proclaimed "clean energy", "renewable" and "environmentally friendly" marketing of energy infrastructure in reality invisibilizes the extractive, processing, manufacturing and transportation components as well as the various labour regimes that underline the acceptance—or consultation—procedures that legitimize energy extraction infrastructure. Not only should we consider renewable energy as fossil fuel+ technologies, ***but every renewable energy acceptance study should account for the acceptance of the mines, labour and transportation regimes necessary for so-called renewable energy development and energy transition.***

Organizing Delusion: Centuries of "Progress"

With the rise of the green economy and new climate change legislations, renewable energy has come to include the harnessing of wind, solar and other apparently infinite "natural resources" to meet energy consumption needs on an unprecedented and ever-expanding scale (see EIA, 2019). Contrary to the claims of its proponents, it is by no means adequate to address the real problem posed by ecological catastrophe and current levels of energy consumption. Not only do "renewable energy" systems require the manufacturing of various excavation machinery, mining and transportation equipment (based on hydrocarbons and extraction), but they also entail the construction of energy-intensive and toxic-waste-producing mineral-processing facilities. Meanwhile, policy debates and mitigation practices focus on the technocratic issue of energy consumption, often leaving unchallenged the political-economic violence intrinsic to organizing and maintaining national and inter-linked global markets.

Shifting industrial society's production and consumption patterns to so-called renewable energy does nothing to remake the exploitative relationships with the Earth and its ecosystems that ***industrialized humans***—people acclimated to and dependent upon an industrial capitalist way of life—have created and reproduce. The ideology of progress reigns supreme within the "Left" and "Right", which, aided by the myth of human supremacy (Springer et al., 2020), justifies the systemic sacrifice of human and non-human life to the beat of techno-capitalist progress. Techno-capitalist systems are far removed from perpetuating ecological restoration and harmony (see Dunlap & Sullivan, 2020), instead opting for the proliferation of industrial and computational infrastructures (Dunlap & Jakobsen, 2020). The excessive concern with possible energy solutions within capitalism— as opposed to more fundamental social transformations— demonstrates peoples' inability to imagine any other way of living, blinding people to the need for a deeper socio-ecological insurrection that climate change, mass extinction and ecological toxification has made necessary. Even "strong community energy" (Devine-Wright, 2019) and "energy democracy" (Batel, 2020; Naumann & Rudolph, 2020)—with their undeniable developmental benefits—are still implicated in a supply chain of ecological destruction and exploitative violence. Even if these violent extractivist relationships have naturalized in market-societies, the habitual patterns of industrial society should be challenged and minimized to the best of peoples' abilities.

Situating "Renewable" Extractive Violence

In 1980, American Indian Movement (AIM) activist Russell Means explained the uncomfortable reality of extractivism in Native territories. Confronting a room of Marxists and revolutionary Communists about their desire for industrialism, Means (1985, p. 25) exclaimed:

> Right now, today, we who live on the Pine Ridge Reservation are living in what Euro society has designated a "national sacrifice area." What this means is that we have a lot of uranium deposits here and Euro culture (not us) needs this uranium as energy production material. The cheapest, most efficient way for industry to extract and deal with the processing of this uranium is to dump the waste byproducts right here at the digging sites. Right here where we live. This waste is radioactive and will make the entire region uninhabitable forever. This is considered by industry, and the white society which created this industry, to be an "acceptable" price to pay for energy resource development. Along the way they also plan to drain the water-table under this area of South Dakota as part of the industrial process, so the region becomes doubly uninhabitable. The same sort of thing is happening down in the land of the Navajo and Hopi, up in the land of the Northern Cheyenne and Crow, and elsewhere. Over 60 percent of all U.S. energy resources have been found to lie under reservation land, so there's no way this can be called a minor issue. For American Indians it's a question of survival in the purest sense of the term. For white society and its industry, it's a question of being able to continue to exist in their present form.
>
> We are resisting being turned into a national sacrifice area. We're resisting being turned into a national sacrifice people. The costs of this industrial process are not acceptable to us. It is genocide to dig the uranium here and to drain the water-table, no more, no less. So the reasons for our resistance are obvious enough and shouldn't have to be explained further. To anyone.

As with the mining of fossil fuels, the mining of raw materials, siting and implementation of renewable energy systems also entails the creation of sacrifice zones, often on Indigenous land. The extractive costs of renewables, associated with the "genocide-ecocide nexus" (see Short, 2016; Dunlap, 2018, 2020c; Normann, 2020), are not accounted for in the social acceptance of wind, solar and other energy systems.

These projects have thus been confronted with considerable pushback from rural and Indigenous populations (see Avila, 2018; Dunlap, 2019a; Finley-Brook & Thomas, 2011; Franquesa, 2018; Lawrence, 2014;

Normann, 2020; Siamanta, 2019) and the struggles against extraction outlined by Means have only continued to intensify overtime. By clinging to ideas like "sustainable development" and the "green economy", progressives and other conscientious citizens are staking the future of the planet on dubious mechanisms of oversight, rife with ideological and material conflicts of interest (Menton et al., 2020). The proliferation of voluntary UN standards, corporate social responsibility initiatives, private auditing firms and free, prior and informed consent (FPIC) processes (see Dunlap, 2018, 2019a; Schilling-Vacaflor & Flemmer, 2019) are well-intentioned "Band-Aid solutions". One can consider them as coercive pacification or "acceptance" technologies (see Dunlap, 2018, 2020b), as they ultimately cover over the real costs of extractivism, especially for the Indigenous and rural people most affected by it. "Are local benefits only a form of 'bribery'?" ask Naumann and Rudolph (2020, p. 100) speaking to this dynamic of land control and persuasion. These "social technology" programs incentivize—if not organize—human populations to trade in their ecosystems for "benefits", "development" and techno-capitalist integration, forfeiting other socio-ecological or post-development and degrowth alternatives (Kothari et al., 2019). Moreover, explain Lina Àlvarez and Brendan Coolsaet (2020, p. 57), calling for "the distribution of environmental impacts", characteristic of liberal inclusionary participation or self-regulation schemes, "implies that nature can be objectified, exploited and turned into a distributable good" even if the idea is "incompatible with Indigenous, Afro-Colombian, and Indian peasant modes of life".

The distinctions drawn between fossil fuels and "renewable energy" involve a sleight of hand that masks the continued socio-ecological degradations necessary for the continuation of consumer society and its ecological modernization (Bonds & Downey, 2012). Renewable energy requires immense amounts of mineral and fossil fuel resources, both in the construction of machinery necessary for extraction and for the manufacturing, transportation, construction and operation of industrial-scale "renewable energy" systems.

Energy infrastructure, whether "green" (e.g. wind turbines, solar panels, power lines) or conventional (e.g. mines, hydraulic fracturing, oil pipelines), are considered "critical infrastructure" of national security importance (Crosby & Monaghan, 2018; Dunlap, 2020a). The status of critical infrastructure and strategic investments renders the opponents of energy infrastructural development targets of political repression. There are numerous accounts of this repression, from protesting pipelines in Unist'ot'em and Wet'suwet'en territories in North America (Crosby & Monaghan, 2018) to wind energy development in Oaxaca (Boyer, 2019; Dunlap, 2019a; Howe, 2019) or energy transformer construction in France (Dunlap, 2020b). Depending on the socio-political context, the police, military or extra-judicial security forces are trained in the tactics of counterinsurgency and employ these repressive tools to manage the conflict in favour of the state, its economic interests and/or private partners. Counterinsurgency is well documented in sites of natural resource extraction. It combines the brute force of "hard" conventional warfare—such as shootings, riot control, rape as a tactic of pacification[2] and so on—and "soft" social warfare strategies that form a larger mutually reinforcing governmental-corporate strategy designed to discipline, enchant and engineer the "hearts" and "minds" of target populations (see Dunlap, 2020a). By initiating "economic and social activities" related to opening roads, waterways and information technology (e.g. radios, telephones, computers, solar panels), privatizing and titling land as well as widening existing politico-ethnic divisions remain important internal processes of social pacification and extraction strategies (Dunlap, 2020a). These frequently include the pre-emptive and systematic targeting of non-violent protesters to enforce the present trajectory of capitalist political economy. For all these reasons, industrial and utility-scale "renewable energy" are more accurately called "Fossil Fuel+" systems. The + indicates the added energy production of the "renewable" component, or "force multiplier", present in these systems.

The Extractive Cost of "Renewables"

Let us focus the discussion on wind energy, the poster child of "renewable energy". Wind energy is increasingly becoming the preeminent approach to climate change mitigation (Naumann & Rudolph, 2020). While there has been a lot of work done about site implementation (Batel, 2020; Boyer, 2019; Creamer et al., 2019; Devine-Wright, 2019; Dunlap, 2019a; Franquesa, 2018; Howe, 2019; Normann, 2020; Siamanta, 2017, 2019), the extractive costs remain under acknowledged.

Consider, for example, the resources required to construct a single two-megawatt wind turbine. One of these turbines uses roughly 150 metric tons of steel for reinforced concrete foundations, 250 metric tons for the rotor hubs and nacelles and 500 metric tons for the tower (Smil, 2016a).

Each turbine also roughly includes 3.6 tons of copper per megawatt (Smith, 2014). Furthermore, industrial steel production is impossible without burning coal, as metallurgical coal—or coking coal—is a vital ingredient in the process (Dıez et al., 2002; Smil, 2016b).[3] Drawing on World Bank report (Arrobas et al., 2017), Jason Hickel (2019, n.p.) estimates the raw materials for seven terawatts from wind and solar infrastructure by 2050 will require mining "34 million metrics tons of copper, 40 million tons of lead, 50 million tons of zinc, 162 tons of aluminium, and no less than 4.8 billion tons of iron". Clean energy transition "is expected to be much more mineral intensive than fossil-fuel based electricity generation", explains a recent World Bank report (Hund et al., 2020, p. 37, 71), and this material intensity "increases with the level of decarbonisation". World Bank reports not only exclude secondary energy infrastructures (e.g. transformers, transmission lines), but also do not take into account the hydrocarbon fuel extraction necessary for mining, processing, manufacturing and transporting raw materials and manufactured components.

Now, imagine regions like the Isthmus of Tehuantepec, where over 2000 wind turbines operate to provide energy to companies like Walmart and Grupo Bimbo as well as industrial construction and mining companies. (Boyer, 2019; Dunlap, 2019a; Howe, 2019). These turbines clearly require significant amounts of prior mining in order to operate. But *every stage* of the mining process, from extraction, processing, manufacturing, transport, construction, to some degree requires a large expenditure of fossil fuels, a fact that is often neglected or minimized in the ecological accounting of wind energy. According to Guezuraga et al. (2012, pp. 40-1), the main consumers of energy and producers of CO_2 for turbines are "the production of stainless steel, followed by concrete and cast iron", while "plastic production represents the most energy intensive process of all materials". While from the perspective of carbon accounting, steel, concrete and cast iron production are the main consumers of energy, the ecological cost of mining and processing rare earth minerals to create permanent magnet generators or "vitamin" upgrades into steels[4] in wind turbines remains publicly neglected.

Where do these minerals come from, and what is the ecological cost of their extraction? Many of the rare earth minerals required for the operation of the turbines—such as dysprosium, praseodymium, neodymium and terbium—come from places like Baotou, Inner Mongolia, and Ganzhou, South East China. Around 85–98% of rare earth minerals used in wind turbines, electric cars, smartphones and other technologies were produced in these locations between the late 1980s and 2015 (Hongiao, 2016; Raman, 2013).

A BBC report (Maughan, 2015) calls the Baotou mining and processing area "hell on Earth", a terrifying, dystopian industrial environment filled with pollution and cluttered with factories, pipelines, high-tension wires and artificial lakes oozing "black, barely-liquid, toxic sludge" that "tested at around three times background radiation (Maughan, 2015, n.p.)". Containing "two hundred million tons of radioactive slurry" (Klinger, 2017, p. 120), the lake is located thirteen square kilometres north of the Yellow River and twelve kilometres from Baotou city centre and has notable "structural vulnerabilities". The lake containing the "bulk of the radioactivity is left behind in the slurry, which continues to generate radon gas", explains Julie Klinger (2017, p. 120), and retains as much as 85% of the radioactivity after the ore is removed. The mine and processing area, Klinger (2017) also observes, hosts wind turbines on lands agro-pastoral nomads struggle to maintain.

Rare earth mining is also disastrously inefficient. Mined with open pit, underground or leached in-situ methods, rare earth ore deposits contain "low concentrations [of desired minerals] ranging from 10 to a few hundred parts per million by weight" (Yang et al., 2013, p. 133). Most concerning, according to Nawshad Haque et al. (2014, p. 621), is that "[t]he mining and processing steps for refining of rare earths tend to be energy, water and chemical intensive with significant environment risks affecting water discharges (radionuclides, mainly thorium and uranium; heavy metals; acid; fluorides), tailing management and air emissions". In the Baotou district, "the three leading causes of death", explains Klinger (2017, p. 121), are cancer; unspecified poisoning and accidents; and infant mortality", which includes a threefold cancer increase between 1990 and 1997 (in an area where there are already high rates compared to the national average). In order to make 2050 "renewable energy" benchmarks, rare earth mineral extraction, Hickel (2019) estimates, will rise between 35% and 70%, which will increase further with the projected fleet of two billion electric vehicles, adding another 70% increase in extraction for neodymium and dysprosium. Meanwhile 34 million metrics tons of copper extraction will double, while cobalt extraction will "increase by a factor of almost four" (Hickel, 2019, n.p.). Mineral extraction is a serious geopolitical ecology issue that is publicly ignored, yet central to energy security and environmental policy.

Wind energy thus involves socially and ecologically destructive mining processes that produce large amounts of mining tailings (or waste) containing heavy metals, thorium and radioactive materials that go into the air, water, soil, animals and people's bodies (see Klinger, 2017). This also includes the waste produced by so-called renewables. "3.1 MW wind turbine created 772 to 1807 tons of landfill waste, 40 to 85 tons of waste sent for incineration and about 7.3 tons of e-waste per unit", explain Benjamin Sovacool et al. (2020: 4), who estimate that 100,000 new wind turbines by 2050 to meet climate mitigation standards "will result in another 730,000 tons of e-waste". Recycling capacities are low, varying between materials, and still retain roughly a 20% recycling rate (Sovacool et al., 2020, p. 2; Habib & Wenzel, 2014), which the EU is currently trying to improve. The quantity and intensity of "upstream" and "downstream" pollution is difficult to measure, not only for political but also epistemic reasons, making full accounting for all ecosystem impacts not only costly but also impossible to achieve. Resource extraction, infrastructural development and corresponding waste outputs establish a colonial continuity, meanwhile spreading rippling and multi-sited ecocidal effects.

Amory Lovins (2017) points out that, while in theory wind turbines could be built without permanent magnets (geared turbines), they are still prevalent in utility-scale wind parks—especially those located offshore or in areas with extreme wind conditions. The increased use and proliferation of geared wind turbines will result in increased resource extractivism (copper, iron ore, cobalt, aluminium etc.), industrial modes of production (to create and maintain mineral processing facilities). Like other industrial enchantments (such as computers or smart technologies), wind farms continue to require levels of extraction that generate toxic and radioactive waste that do not show up on carbon accounting sheets and outdated life cycle assessments (Kiezebrink et al., 2018).[5] The green economy, in fact, is expanding its demand for destructive mining of iron ore, copper, oil, cobalt and rare earth minerals, but also contributing and expanding sacrifice areas. Climate change mitigation must question the present mode of capitalist development, and industrialism in general, if there is any hope of repairing ecosystems and (gradually) stabilizing erratic weather patterns and climatic changes.

Sustainable Violence: There Is no Such Thing as Renewable Energy

The political and environmental costs of implementing these fossil fuel+ systems are high. Scale, placement, mitigation practices, energy-use and decommissioning are foundational for assessing the viability and long-term socio-ecological sustainability of wind turbines. This means taking cognizance of the quantity and location of large-scale turbines, as well as the various political and socio-geographical factors involved in their construction (see Naumann & Rudolph, 2020), if one aims to repair socio-ecological relations and de-grow the existing techno-capitalist system.

It is ill-advised to place wind turbines in areas with fresh groundwater, on lands used by semi-subsistence Indigenous groups, within 1.5 kilometres of people's homes, or on farming and fishing territories. This is precisely what has happened on the Isthmus of Tehuantepec (Dunlap, 2019a). The construction and placement of wind turbines requires the creation of roads that clear trees, destroy animal habitats and compact soil. They also require the creation of wind turbine foundations that range, depending on the site, between 7 and 14 meters (23–45 ft.) deep and about 16–21 meters (52–68 ft.) in diameter. The foundations require the filling of ground water with solidifying chemicals followed by steel-reinforced concrete. Then during operation, leaking oil seeps into the ground where animals graze and into water wells where people drink (Dunlap, 2019a). And this leaves aside the effects of concrete production, as well as the violence involved in building wind or other renewable energy systems on Indigenous territory. On top of all this, each wind turbine only has roughly a 25- to 35-year lifespan before it needs to be decommissioned and, hopefully, recycled, which is currently done at an unsatisfactory rate below 20% (as mentioned above).

What is all this energy used for? "Strong" community and fossil fuel+ democracy projects tend to be more socially inclusive, to operate on a micro-to-meso scale and to provide energy to local or affected communities. Commercial, or in the UK decentralized "local energy" (Devine-Wright, 2019: 894), fossil fuel+ systems capture, export and privatize vital resources. Large-scale and market-based fossil fuel+ systems are opening and widening new wind, solar and other natural resource frontiers to renew and expand capitalism (Dunlap, 2019a: Franquesa, 2018; Siamanta, 2017, 2019; Normann, 2020), as they spread capitalist relationships and entrench unsustainable practices of production and consumption.

The techniques and technologies that are helping corporations expand in ostensibly "green" directions will be applied to power military infrastructures and equipment. There are numerous accounts of this: solar energy powering on military bases in Iraq (Bunn, 2016),[6] wind energy powering the Mexican army's activities (Dunlap, 2017), aircraft carriers that run on biofuels (Bigger & Neimark, 2017) as well as California police stations upgrading their buildings with solar panels (City of Mill Valley, 2017). These relations support the expansion of capitalism while obscuring its wrenching crises—thereby obstructing effective identification of problems and their reconciliation with people-led or government-led actions. Industry and security forces are beginning to acknowledge their ecologically destructive operations, and repressive forces are looking for ways to be seen as ecologically "sustainable". "Sustainable violence" (Dunlap, 2017) recognizes the deployment of fossil fuel+ technologies in the service of security forces. So-called renewable energy economizes, renews and attempts to make repressive military, police and extra-judicial operations energetically "sustainable". The police, in the US, for example, well-known for their structural racism and political violence (Williams, 2007), are now "greening" and progressively advancing towards articulating systematic acts of "sustainable violence", executing repressive operations with "environmentally friendly" technologies. Fossil fuel+ systems are paving the way for systems of sustainable violence, which attempts to economize repression and use of force to continue the present trajectory of techno-capitalist progress and ecological destruction.

Hydrocarbon industries—whether coal, natural gas or oil—are beginning to invest and use the renewable industry to legitimize their resource extraction operations and diversify their energy-related holdings. Examples range from Gas Natural Fenosa, which is investing in wind parks in Mexico (Dunlap, 2019a); RWE in Germany, operator of the largest coal mine in the country (Brock, 2020), which is setting up their own green daughter company—"Innogy"—to invest in wind energy and other "renewables" after spending years subverting and lobbying against them (Brock & Dunlap, 2018). Grupo Mexico is also buying wind power in Mexico and solar parks in the US to cloak their company in a "green" image, meanwhile powering the extraction of raw materials with renewable sources (Dunlap, 2020a), which is a concerted development strategy in Chile (Furnaro, 2019). The magazine *Energy and Mines* is completely dedicated to promoting the use of renewable energy in mining. Dunlap and Andrea Brock (2021) call this the "renewable energy-extraction nexus" (Dunlap,

2018),[7] which demonstrates the intimate relationship between forms of extraction—whether wind, natural gas, coal or copper—necessary for the acquisition of resources and fossil fuel+ development. The renewable energy-extraction nexus embodies the network of extractive industries and utility providers that make up the body of the capitalist state. This includes the intricate web of subsidization, collaboration and, at times, competition among these entities that renews the spread of capitalist infrastructure across the planet (see Dunlap & Jakobsen, 2020). This expansion happens at a great disregard for the costs involved, whether for people (particularly Indigenous and rural communities in both the Global North and South), animals, plants or geophysical nature. As we know it, there is no such thing as "renewable energy", plain and simple. It is all based on extreme hydrocarbon and mineral extraction, which—as mentioned—relates to oppressive labour and ecological conditions.

All "renewable resources" are a vital or kinetic energetic force stopped or redirected: from the direction of the wind, to the flow of a river, to the beams of sunlight. Infrastructural colonization of landscapes creates an interface: shaping environments, fusing machines to habitats and creating an apparatus to absorb and convert vital energy flows (see Dunlap, 2020). Rivers, wind, sun, tidal wave and other vital forces are captured, domesticated and transformed into "energy" (see Daggett, 2019), no longer serving the flora, fauna and soil, being instead reconfigured to power industrial infrastructure and computational systems that break renewable/reciprocal energy cycles. Ecological economists' attempt to organize "sociometabolic approach to examine the material flows" (Martinez-Alier & Walter, 2016, p. 58), which investigates "the metabolism of modern energy technologies such as combustion engines, hydroelectric generators, nuclear power plants, biofuels, and photovoltaic panels" and more (Hornborg, 2020, p. 10). Widening the economistic calculations of energy use—or the documentation of socio-ecological plunder—this materialist approach is useful, but still qualitatively limited in documenting the profound psycho-social and ontological impacts of energy extractivism and techno-capitalist modes of social organization.

Vital life or flows are usurped from natural cycles and redirected into lifeless infrastructure. This is a fate shared by human, solar and wind "resources". The vital energetic wind, solar, water and other elemental flows make up the self-sustaining life source of the world, and instead of seeing these vital life forces as "gods", "spirits" or lifeways, industrial humans instead reduced them to "kinetic energy" and "renewable energy" to be exploited and injected into industrial-computational infrastructures that propel capitalist processes and institutions. Conventional and fossil fuel+ infrastructures are designed to degrade ecosystems (in one or multiple sites), subjugate people attached to ecosystems and, overall, sacrifice energetic forces to techno-capitalist production. The challenge for energy democracy and "strong" community energy systems—and the ecologically sensitive—is to repair this multi-scalar planetary energetic cycle. Social acceptance takes place on many scales and locations, not only with the site placement of wind turbines and solar panels but the series of preceding mines and manufacturing facilities before it.

Vital life is reconfigured and consumed, and we must ask ourselves: to what ends? Ecologically, socially and economically costly and energy-intensive infrastructures are constructed to interrupt the pathways, or cycles, of vital life to power superstores, office buildings, mineral processing facilities and the overall conquest of both human and nonhuman natures. In short, so-called renewable energy is always taking something from the environment or at best redirecting it to other activities, altering weather patterns and potentially weakening the circular vitality of ecosystems or "earth systems" (Tabassum-Abbasi et al., 2014, pp. 280–81). The "extraction of the wind's kinetic energy" are generating wind "velocity deficits of the order of 10% or more" downstream of onshore and offshore wind factories, which result in "significant warming trends", thereby aggravating climate changes (Abbasi, Tabassum-Abbasi, & Tasneem-Abasi, 2016, p. 1592). These processes are implicitly built on neglecting the sensitive reciprocal ecological relationships between human and non-human. Techno-capitalist-industrial systems break energetic cycles of ecological reciprocation, redirecting energy traditionally used to nourish landscapes, forests, human and non-human populations into extracting energy to destroy and colonize these landscapes and populations in the service of various competing and complementing ideologies of control and mass-consumption. Renewable energy serves as another apparatus to intensify and continue this rupture in ecological reciprocation, redirecting energy away from supporting ecosystems and their inhabitants to producing capitalist infrastructures, natures and humans bent on resource control. "Renewable energy", as it stands, is a lie: there is only fossil fuel+. This impresses the point that until this energetic rupture in the cycles of natural and planetary flows are repaired and rebalanced, then there is no such thing as renewable energy.

Conclusion

The preceding considerations allow us to recognize so-called renewable energy as fossil fuel+ technologies that are ***renewing destruction***. It entails revived and intensified relations of domination that have much in common with colonial and centre-periphery dynamics. Habitats, ecosystems and the humans and non-humans that live in them are subordinated to the order of infrastructural development in order to produce energy for cities, industries and, overall, techno-capitalist growth. When people embrace fossil fuel+ systems, many do not realize that they entail various forms of violence against people, environments and animals. Wrapped in public relations, marketing and "climate youth demands" (see Cory Morningstar, 2019), fossil fuel+ systems demand extreme hydrocarbon and mineral extraction, socio-ecologically abhorrent industrial facilities and labour conditions as well as the production of large amounts of waste. This, for the most part, is only made socially acceptable by being placed out of sight and out of mind—invisibilized—in the materially poor, rural and Indigenous territories of the world.

When liberals, "progressives", "the Left" and even environmental justice activists applaud the large-scale transition to fossil fuel+ technologies, they ignore the many intrinsic hazards that would otherwise be unacceptable to them. Displacing fossil fuel industries to the Global South, where there are fewer environmental regulations and political rights, makes possible the use of extreme forms of state-private security violence against anyone who might protest them. If people do not confront these facts, then the solution of today—like previous energy systems and regime changes—will likely result in the complicated tyrannies of tomorrow. Recognizing "renewable energy" as fossil fuel+ is a first step to combat the eco-modernist myth.

When researchers discuss "the social acceptance of renewable energy technologies" or energy democracy, we need to acknowledge what passes for "acceptance" (e.g. acquiescence, adverse incorporation and "soft" and "hard" pacification of opposition) and apply this analysis to entire supply chains. The so-called acceptance of fossil fuel+ technologies depends on and takes place in multiple sites, numerous mines, smelting facilities and industrial centres, which are not "participatory" in any meaningful way, but are instead subject to civil-military counterinsurgency controls. This raises questions about the social engineering of resource extraction and ecological crises in general (see Dunlap, 2020a; Verweijen & Dunlap, 2021). If people are serious about "renewable energy" and "energy transition", then this process requires ethnographically grounded qualitative and quantitative research into the life cycles of renewable/fossil fuel+ technologies at multiple sites, scales and employing decolonial epistemologies.

At the very least, by relabelling "renewable energy" as fossil fuel+ we create the groundwork for greater environmental considerations and the enactment of radical ecological alternatives to confront consumer society and its marketed solutions. There is no such thing as renewable energy when vital systems are disrupted and redirected into politico-economic systems of mass human and non-human resource extraction, exploitation and consumption. We must ask ourselves how humans can reconnect and develop alongside pre-existing patterns and cycles of renewable energy flows.

Notes

1. Malm's (2016) advocacy for "renewable energy" at the end of his book exemplifies this false dichotomy—demonstrating a common, but also normalized cognitive dissonance concerning (highly toxic) extractive, manufacturing and labour regimes necessary for fossil fuel+ systems.
2. See Power & People. (2012) *Peru: Undermining Justice*. Available at: https://www.aljazeera.com/programmes/peopleandpower/2012/05/20125311829466420.html
3. There is a rise of "green" smelters, using hydropower (see Hobson, 2017). This requires further investigation (look for Susanne Norman's forthcoming work).
4. This refers to the process of adding or splicing rare earth minerals into steel to avoid rusting as well as enhance durability and other infrastructural functions.
5. Personal communication with Judith Pigneur.
6. "The Military is Going Green" with Arnold Schwarzenegger. Available at: http://channel.national geographic.com/years-of-living-dangerously/videos/the-militarys-going-green/
7. The renewable energy-extraction nexus was written in 2017 for a book chapter in *the Anarchist Political Ecology Volume*, which has experienced severe publication delays.

References

Abbasi, S. A., Tabassum-Abbasi, & Tasneem-Abbasi. (2016). Impact of wind-energy generation on climate: A rising spectre. *Renewable and Sustainable Energy Reviews, 1*, 591591–591598. https://doi.org/10.1016/j.rser.2015.12.262

Álvarez, L., & Coolsaet, B. (2020). Decolonizing environmental justice studies. *Capitalism Nature Socialism, 31*(2), 50–69. https://doi.org/10.1080/10455752.2018.1558272

Arrobas, D. L. P., Hund, K. L., McCormick, M. S., Ningthoujam, J., & Drexhage, J. R. (2017). *The growing role of minerals and metals for a low carbon future.* The World Bank. Available at: http://documents.worldbank.org/curated/en/207371500386458722/The-Growing-Role-of-Minerals-and-Metals-for-a-Low-Carbon-Future

Avila, S. (2018). Environmental justice and the expanding geography of wind power conflicts. *Sustainability Science, 13*(3), 599–616. https://doi.org/10.1007/s11625-018-0547-4

Batel, S. (2020). Research on the social acceptance of renewable energy technologies. Past, present and future. *Energy Research & Social Science, 68*, 1–5.

Bigger, P., & Neimark, B. D. (2017). Weaponizing nature. The geopolitical ecology of the US Navy's bioful program. *Political Geography, 60*(1), 13–22. https://doi.org/10.1016/j.polgeo.2017.03.007

Bonds, E., & Downey, L. (2012). Green technology and ecologically unequal exchange. *Journal of World-Systems Research, 18*(2), 167–186.

Boyer, D. (2019). *Energopolitics.* Duke University Press.

Brock, A. (2020). Securing accumulation by restoration. *Environment and Planning E, 1*, 1–21. https://doi.org/10.1177/2514848620924597

Brock, A., & Dunlap, A. (2018). Normalising corporate counterinsurgency: Engineering consent, managing resistance and greening destruction around the Hambach coal mine and beyond. *Political Geography, 62*(1), 33–47.

Bunn, J. (2016). *Schwarzenegger talks Army green energy in Kuwait.* US Army. https://www.army.mil/article/167274/schwarzenegger_talks_army_green_energy_in_kuwait.

Büscherr, B., & Fletcher, R. (2020). *The conservation revolution.* Verso.

City of Mill Valley. (2017). *City to host ribbon cutting for solar project to power community center.* https://www.enjoymillvalley.com/-blog/city-to-host-ribbon-cutting-for-solar-project-to-power-community-center.

Creamer, E., Aiken, G. T., van Veelen, B., Walker, G., & Devine-Wright, P. (2019). Community renewable energy: What does it do? *Energy Research & Social Science, 57*, 1–6. https://doi.org/10.1016/j.erss.2019.101223

Crosby, A. C., & Monaghan, J. (2018). *Policing indigenous movements: Dissent and the security state.* Fernwood Publishing.

Daggett, C. N. (2019). *The birth of energy*. Duke University Press.

Devine-Wright, P. (2019). Community versus local energy in a context of climate emergency. *Nature Energy, 4*(11), 894–896. https://doi.org/10.1038/s41560-019-0459-2

Dıez, M., Alvarez, R., & Barriocanal, C. (2002). Coal for metallurgical coke production. *International Journal of Coal Geology, 50*(1–4), 389–412. https://doi.org/10.1016/S0166-5162(02)00123-4

Dunlap, A. (2017). Wind energy: Toward a "sustainable violence" in Oaxaca, Mexico. *NACLA, 49*(4), 483–488. https://doi.org/10.1080/10714839.2017.1409378

Dunlap, A. (2018). "A bureaucratic trap:" free, prior and informed consent (FPIC) and wind energy development in Juchitán, Mexico. *Capitalism Nature Socialism, 29*(4), 88–108. https://doi.org/10.1080/10455752.2017.1334219

Dunlap, A. (2019a). *Renewing destruction: Wind energy development, conflict and resistance in a Latin American context*. Rowman & Littlefield.

Dunlap, A. (2019b). Green New Deal Part II. *Terra Nullius*. https://www.sum.uio.no/forskning/blogg/terra-nullius/green-new-deal-part-II-good-bad-and-the-ugly.html.

Dunlap, A. (2020a). Wind, coal, and copper: The politics of land grabbing, counterinsurgency, and the social engineering of extraction. *Globalizations, 17*(4), 661–682. https://doi.org/10.1080/14747731.2019.1682789

Dunlap, A. (2020b). Bureaucratic land grabbing for infrastructural colonization: Renewable energy, L'Amassada and resistance in southern France. *Human Geography, 13*(2), 1–22.

Dunlap, A. (2020c). The Politics of Ecocide, Genocide and Megaprojects: Interrogating Natural Resource Extraction, Identity and the Normalization of Erasure. *Journal of Genocide Research*, 1–26. https://doi.org/10.1080/14623528.2020.1754051

Dunlap, A., & Brock, A. (2021). When the Wolf Guards the Sheep: Green Extractivism in Germany and Mexico. In S. Springer, M. Locret, J. Mateer, et al. (Eds.), *Anarchist Political Ecology Volume 3* (pp. 1–20). London: Rowman & Littlefield.

Dunlap, A., & Jakobsen, J. (2020). *The violent Technologies of Extraction*. Palgrave.

Dunlap, A., & Sullivan, S. (2020). A faultline in neoliberal environmental governance scholarship? Or, why accumulation-by-alienation matters. *Environment and Planning E, 3*(2), 552–579. https://doi.org/10.1177/2514848619874691

EIA. (2019). *EIA projects nearly 50% increase in world energy usage by 2050, led by growth in Asia*. US Energy Information Administration. https://www.eia.gov/todayinenergy/detail.php?id=41433.

Finley-Brook, M., & Thomas, C. (2011). Renewable energy and human rights violations: Illustrative cases from indigenous territories in Panama. *Annals of the Association of American Geographers, 101*(4), 863–872. https://doi.org/10.1080/00045608.2011.568873

Franquesa, J. (2018). *Power struggles: Dignity, value, and the renewable energy frontier in Spain.* Indiana University Press.

Furnaro, A. (2019). Neoliberal energy transition. *Environment and Planning E, 1*, 1–25. https://doi.org/10.1177/2514848619874685

Guezuraga, B., Zauner, R., & Pölz, W. (2012). Life cycle assessment of two different 2 MW class wind turbines. *Renewable Energy, 37*(1), 37–44. https://doi.org/10.1016/j.renene.2011.05.008

Habib, K., & Wenzel, H. (2014). Exploring rare earth supply constraints for the emerging clean energy technologies and the role of recycling. *Journal of Cleaner Production, 84*(1), 348–359. https://doi.org/10.1016/j.jclepro.2014.04.035

Haque, N., Hughes, A., Lim, S., & Vernon, C. (2014). Rare earth elements: Overview of mining, mineralogy, uses, sustainability and environmental impact. *Resources, 3*(4), 614–635. https://doi.org/10.3390/resources3040614

Hickel, J. (2019). The limits of clean energy. *Foreign Policy*. https://foreignpolicy.com/2019/09/06/the-path-to-clean-energy-will-be-very-dirty-climate-change-renewables/.

Hobson, P. (2017). Hydro-powered smelters charge premium prices for 'green' aluminum. Reuters, https://www.reuters.com/article/us-aluminium-sales-environment/hydro-powered-smelters-charge-premium-prices-for-green-aluminum-idUSKBN1AI1CF.

Hongiao, L. (2016). *The bottleneck of a low-carbon future. China dialogue*, https://chinadialogue.net/article/9209-The-bottleneck-of-a-low-carbon-future.

Hornborg, A. (2020). Energy, space, and movement: Toward a framework for theorizing energy justice. *Geografiska Annaler: Series B, Human Geography, 102*(1), 8–20. https://doi.org/10.1080/04353684.2019.1682939

Howe, C. (2019). *Ecologics*. Duke University Press.

Kiezebrink, V., Wilde-Ramsing, J., & Kate Gt. (2018). *Human rights in wind turbine supply chains. SOMO and Actionaid*. https://www.somo.nl/wp-content/uploads/2018/01/Final-ActionAid_Report-Human-Rights-in-Wind-Turbine-Supply-Chains.pdf.

Kothari, A., Salleh, A., Escobar, A., et al. (2019). *Pluriverse: A Post-Development Dictionary*. Delhi: University of Colombia Press.

Hund, K., La Porta, D., Fabregas T.P., Laing, T., & Drexhage J. (2020). *Minerals for Climate Action*. The World Bank Group, http://pubdocs.worldbank.org/en/961711588875536384/Minerals-for-Climate-Action-The-Mineral-Intensity-of-the-Clean-Energy-Transition.pdf.

Klinger, J. M. (2017). *Rare earth frontiers: From terrestrial subsoils to lunar landscapes*. Cornell University Press.

Lawrence, R. (2014). Wind power developments on traditional Saami lands. *Environment and Planning D: Society and Space, 32*(6), 1036–1053. https://doi.org/10.1068/d9012

Lovins, A. (2017). *Clean energy and rare earths: Why not to worry.* Bulletin of the Atomic Scientists, https://thebulletin.org/clean-energy-and-rare-earths-why-not-worry10785.

Malm, A. (2016). *Fossil capital.* Verso Books.

Martinez-Alier, J., & Walter, M. (2016). Social metabolism and conflicts over Extractivism. In F. De Castro, B. Hogenboom, & M. Baud (Eds.), *Environmental governance in Latin America.* Palgrave Macmillan.

Maughan, T. (2015). *The Dystopian lake filled by the world's tech lust.* BBC, http://www.bbc.com/future/story/20150402-the-worst-place-on-earth.

Means, R. (1985). The Same Old Song. In W. Churchill (Ed.), *Marxism and Native Americans* (pp. 19–34). Boston: South End Press.

Menton, M., Larrea, C., Latorre, S., Martinez-Allier, J., Peck, M., Temper, L., & Walter, M. (2020). Environmental justice and the SDGs: From synergies to gaps and contradictions. *Sustainability Science, 1*, 1–16. https://doi.org/10.1007/s11625-020-00789-8

Morningstar, C. (2019). *The Manufacturing of Greta Thunberg—for Consent has been written in six acts.* The Wrong Kind of Green, http://www.wrongkindofgreen.org/2019/01/17/the-manufacturing-of-greta-thunberg-for-consent-the-political-economy-of-the-non-profit-industrial-complex/.

Naumann, M., & Rudolph, D. (2020). Conceptualizing rural energy transitions. Energizing rural studies, ruralizing energy research. *Journal of Rural Studies, 1*, 7397–7104. https://doi.org/10.1016/j.jrurstud.2019.12.011

Normann, S. (2020). Green colonialism in the Nordic context: Exploring southern Saami representations of wind energy development. *Journal of Community Psychology, 1*, 1–18. https://doi.org/10.1002/jcop.22422

Raman, S. (2013). Fossilizing renewable energies. *Science as Culture, 22*, 172–180

Sanders, B. (2019). The Green New Deal. *Bernie Sanders-Official Campaign Website.* https://berniesanders.com/en/issues/green-new-deal.

Schilling-Vacaflor, A., & Flemmer, R. (2019). Mobilising free, prior and informed consent (fpic) from below: A typology of indigenous peoples' agency. *International journal on minority and group rights, 1(aop)*, 1–23. https://doi.org/10.1163/15718115-02702008

Short, D. (2016). *Redefining Genocide: Settler Colonialism, Social Death and Ecocide.* London: Zed Books.

Siamanta, Z. C. (2017). Building a green economy of low carbon: The Greek post-crisis experience of photovoltaics and financial 'green grabbing'. *Journal of Political Ecology, 24*, 258–276. https://doi.org/10.2458/v24i1.20806

Siamanta, Z. C. (2019). Wind parks in post-crisis Greece. *Environment and Planning E, 2*(2), 274–303. https://doi.org/10.1177/2514848619835156

Smil, V. (2016a). *To Get Wind Power You Need Oil.* IEEE, http://spectrum.ieee.org/energy/renewables/to-get-wind-power-you-need-oil.

Smil, V. (2016b). *Energy transitions.* Praeger.

Smith, P. (2014). *Soaring copper prices drive wind farm crime.* Wind Power Monthly: http://www.windpowermonthly.com/article/1281864/soaring-copper-prices-drive-wind-farm-crime.

Sovacool, B. K., Hook, A., Martiskainen, M., Brock, A., & Turnheim, B. (2020). The decarbonisation divide: Contextualizing landscapes of low-carbon exploitation and toxicity in Africa. *Global Environmental Change, 60*, 1–19. https://doi.org/10.1016/j.gloenvcha.2019.102028

Springer, S., Locret, M., & Mateer, J. (2020). *Anarchist political ecology.* Rowman & Littlefield.

Tabassum-Abbasi, P. M., Abbasi, T., & Abbasi, S. A. (2014). Wind energy: Increasing deployment, rising environmental concerns. *Renewable and Sustainable Energy Reviews, 31*(1), 270–288. https://doi.org/10.1016/j.rser.2013.11.019

Verweijen, J., & Dunlap, A. (2021). The evolving techniques of social engineering, land control and managing protest against extractivism: Introducing political (re)actions 'from above'. *Political Geography, 83*, 1–9. https://doi.org/10.1016/j.polgeo.2021.102342Get

Williams, K. (2007 [2004]). *Our enemies in blue: Police and power in America.* South End Press.

Yang, X. J., Lin, A., Li, X.-L., Wu, Y., Zhou, W., & Chen, Z. (2013). China's ion-adsorption rare earth resources, mining consequences and preservation. *Environmental Development, 8*, 131–136. https://doi.org/10.1016/j.envdev.2013.03.006

PART IV

For Interdisciplinarity

CHAPTER 6

ANT Perspective on Wind Power Planning and Social Acceptance—A Call for Interdisciplinarity

Julia Kirch Kirkegaard and Sophie Nyborg

Introduction: Opening Black-Boxed 'Facts' of Landscapes, Resources and Stakeholders

How do some landscapes qualify as appropriate for wind power development and others not? How has the wind, blowing briskly or softly and shaping landscapes and cultures over centuries, become a commodified and valued resource to be capitalised on by some and not by others? How are stakeholders and their (il)legitimate stakes demarcated, and by whom? What are the effects of concepts, categorisations and notions as 'sites and landscapes', 'resources' and 'stakeholders'?

The social acceptance literature has provided rich insights into reasons for local opposition to renewable energy technologies (RET) developments (e.g. Aitken, 2010; Devine-Wright & Howes, 2010; Ellis & Ferraro, 2016; Walker et al., 2010; Wolsink, 2007; Wüstenhagen et al., 2007).

J. K. Kirkegaard (✉) • S. Nyborg
Technical University of Denmark, Roskilde, Denmark
e-mail: jukk@dtu.dk; sonyb@dtu.dk

S. Batel, D. Rudolph (eds.), *A critical approach to the social acceptance of renewable energy infrastructures*,
https://doi.org/10.1007/978-3-030-73699-6_6

Meanwhile, the literature tends to treat some concepts and notions, such as wind farm sites, wind resources and stakeholders, as given 'facts'. We argue that to engage in a critical inquiry into social opposition and acceptance of wind power (and other RET's), there is much to be gained from cross-fertilising the Social Acceptance (SA) literature with that of Actor-Network Theory (ANT) within Science & Technology Studies (STS).

Appropriating the engineering term of 'black box', ANT seeks to critically investigate the socio-material work and tools that have been employed to black box something as a 'matter of fact' (Latour, 2004), in order to reveal the socio-material construction and inherent politics of black boxes and the matters of concern they produce. By 'black boxing' we mean to 'frame', that is, to simplify, bracket and categorise. Black boxing and framing is necessary to assemble any network or socio-technical (market) assemblage into a (temporarily stabilised) whole. Without black boxes, it would be difficult to act or make decisions, for example, on the planning and development of wind farms. Of course, black boxes also include some things within the frame, while excluding other things outside the framing of the black box. Framing and their black boxes are thus inherently fragile, as they entail power structures (Caliskan & Callon, 2010, p. 12, 18), which may be contested, and thus result in unforeseen consequences (so-called overflowing, or what mainstream economics would call 'externalities') such as *trials of strength* (Callon, 1998, p. 33, 252-4). It is the 'facts' contained in the black box that are put under critical inquiry in an ANT analysis, often revealing the enormous amount of socio-technical work that has been required to assemble and maintain something as a 'fact'. It is often found that this requires the employment of calculative devices (Callon, 1991, Callon, 1998; Callon & Muniesa, 2005) that help to associate the black box with certain values. When opening the black box, it becomes possible to see how some of the 'values' are contested, causing controversy.

In this chapter, we provide three empirical snapshots of what the perspective of the socio-material, symmetrical and relational ANT lens might have to offer to the literature on the social acceptance of RET. In particular, we look at the sub-strands *of the anthropology of markets* (e.g. Caliskan & Callon, 2010; Callon, 1998; Finch et al., 2017; Geiger et al., 2014; Geiger & Gross, 2018; Kirkegaard, 2019) and the *sociology of valuation* (e.g. Dussauge et al., 2015; Geiger & Gross, 2018; Kornberger et al., 2015; Roscoe & Townley, 2016). The snapshots serve as examples of three central, though often taken-for-granted (black-boxed), notions and categorisations of (1) wind farm sites, (2) wind resources and (3) stakeholders. Based on the three empirical snapshots, the chapter calls for cross-fertilisation between the two fields. In doing so, we align with several scholars in the field (e.g. Hess & Sovacool, 2020; Labussiere & Nadai, 2018a, b; Batel, 2018; Jolivet & Heiskanen, 2010; Chilvers et al., 2018; Chilvers & Longhurst, 2016; Aaen et al., 2016; Kirkegaard et al., 2020). Most recently, Hess and Sovacool (2020) have used the expression "socio-technical matters" (p. 2), arguing that STS in energy social science can help to inquire into "matters that are both social and technical and that otherwise might be consigned to the black box of uninspected exogenous factors" (Hess & Sovacool, 2020, p. 2). What ANT helps to bring to the fore is the often black-boxed material aspects of wind energy acceptance and the hidden power and valuation struggles that black boxes produce. In other words, the ANT lens can help to shed light on issues of *emergence* (of wind farms, wind power markets and more), *agency* (of heterogeneous actors) and *dynamics* of transformation (ubiquity of overflowing power struggles such as local opposition).

Snapshot 1: The Making of a Landscape and a Site

First, we zoom in on the socio-material process of black boxing and identify certain places as 'wind farm landscapes' and how they are being constructed as 'sites'. The question of whether a landscape qualifies as appropriate for wind turbines is a core issue in the development and deployment of wind power today. The environmental impact assessment (EIA) report is a central framing device in any siting process because the EIA process is aimed at evaluating the environmental consequences (positive or negative) of a proposed plan before any decision is made to move forward with it. Landscape descriptions, zoning and categorisations are a substantial part of an EIA for a planned wind farm. For instance, the landscape needs to be evaluated as conducive to allowing the wind to flow; that is, it needs to be sparsely populated, the soil needs to be suitable for bearing wind turbine foundations, ground water levels should be appropriately low and the landscape should not be host to endangered animals or plant species. Moreover, the EIAs not only build on engineering expertise and 'facts about nature'. The aesthetic suitability of landscapes for wind turbines is often also (e)valuated by experts such as landscape architects who assess whether the landscape composition is such that it can visually accommodate wind turbines and hence, can be transformed from a landscape to a manageable, suitable wind turbine 'site' which developers can exploit.

It is obviously not new to the social acceptance literature, geography and other disciplines, to be interested in sites and landscapes—or in general in the 'locations' of RET projects and their disruption to place attachment (Devine-Wright, 2011). In fact, the advent of distributed RET's has spurred "a new interest in the landscape-energy relationship" to explore the co-development between RET and landscapes (Nadaï & van der Horst, 2010, p. 143), the social construction of nature (Woods, 2003) as well as how the contextually embedded qualities of landscape are represented to construct 'acceptable locations' (Cowell, 2010).

What such studies hint at is that 'appropriate landscapes' as described above are not just out there waiting to be described in a fact-based EIA. Instead these 'landscapes', 'locations', backyards' (Devine-Wright, 2011) and 'places' are constructed and made into sites that fit wind farms just as much as wind farms are made to fit the landscape. Yet, what the SA literature does not pay concerted attention to are the power issues and inclusion and exclusion activities that inherently lie in the classifications and evaluations of landscapes and sites, in, for example, EIAs and how the *material entities* such as calculative devices, soil, water, wind and birds (Nadaï & Labussiere, 2010) play a core role in shaping and stabilising these 'social' (socio-material) power relations and vice-versa. Landscapes and wind turbine sites are indeed political beings, as noted by Labussiere and Nadaï (eds., 2018b), who have recently called for more focus on how 'earth and soil matter' (2018) in the transition towards more RET systems (Latour, 2017).

A few other studies have used an ANT lens to open the black box of appropriate landscapes in order to articulate the politics of non-human actors, and to understand how these are 'made to work', when attempting to understand controversies related to wind turbine planning and development. Krauss (2010), for instance, argues that the rise of wind energy in a coastal region of Northern Germany "is not merely the result of (trans-) national governance strategies, but has to be analysed as a collective effort of the assemblies of people and things that make up this coastal landscape" (Krauss, 2010, p. 196). In the quest to understand the governance of landscapes and their attributes and entities, Kropp (2017) also appreciates a constructivist perspective on the 'production of space' and argues, following Krauss (2010), that 'the whole network of people and things' is brought into focus, thus re-politicising the connections (p. 563). Drawing on Latour's discussion of *Dingpolitik* (Latour, 2005a), Kropp argues that "landscapes, then, are not to be considered as external scenery 'out there' or as spatial containers for external objects, but instead as *cosmopolitical* 'things' that bring people together through their heterogeneous composition and their long-term social relevance" (2018, p. 563). Thus, controversies over landscapes that are enrolled in RET development "include disputes over the most suitable techniques and organisations, sitings, responsibilities and competencies, [and] over who can make decisions about long-lasting socio-technical changes to common spaces" (ibid.).[1] The aim of an ANT analysis is not to point to the rhetorical or political dimensions of certain discursive framings, but to pay close attention to and unpack these disputes. It is about tracing the role that *both* humans and things, calculations, models and so on have in rendering certain politics as given facts (e.g. the suitability of a landscape), and how these black-boxed facts grant power to some actors over others in attempts to settle the disputes.

In ongoing research we also trace the network of people and things enrolled in attempts to build landscapes and sites by using the case of a wind farm in the north-western part of Denmark.[2] Here, the meadow landscape has been described by municipal planners and EIA consultants, as well as developers, as ideal to host a large wind turbine park. However, we argue that this landscape was actually shaped to become so (see also Krauss, 2010) through socio-material work of black boxing/framing. That is, the landscape has been artificially shaped through several phases over the last couple of centuries (e.g. through the dikes that are draining areas, removing of hedgerows in agricultural fields, engaging in industrial farming) and employing planning tools and zoning that allows certain developments of an area to happen, that is, that the area will be used for farming instead of wildlife or urban development and so on. Through an ANT lens, it becomes possible to unpack the framing and (monetary and non-monetary) valuation exercises that have been done in connection with 'the site' to make it fit to the imagined wind farm. Although the facts, categories and maps that this siting process was built upon are black boxes, the study also shows that these always leak, causing local opposition. The socio-material framing processes of landscape-shaping have been an underlying, though often overseen, reason for much of the local contestation and 'non-acceptance'. Thus, the built or cultivated environment, for example, agricultural fields, dammed coastal areas and so on, has vested interests built into it. Even though it is often said that wind farms are 'made to fit to the landscape' it also happens the other way round: landscapes are made to fit wind farms.

Snapshot 2: The Making of an Energy Resource

Next, we look into how wind resources are the result of socio-material work of black boxing. What does it take to transform the blowing wind into a valuable 'resource' that certain actors can commodify and earn profit from? The socio-technical work of framing and valuation required to transform RET "from untamed, heterogeneous, or difficult-to-access forms of energy" (Nadaï & Labussiere, 2018: 57) to the production and marketisation of wind power also produces certain inclusion and exclusion effects that easily spur power struggles (Labussiere & Nadaï (eds.), 2018b; Kirkegaard et al., 2020; Kirkegaard, 2019; Brighenti, 2018).

Black boxing/framing is needed in order to make the wind into an interesting investment object. That is, before any 'investor' (be it a wind turbine manufacturer, wind farm developer, a national government, a municipal council etc.) decides on investing time, money and/or efforts in the wind power sector, the investor must be able to calculate/valuate or predict the outcome or profits that may accrue from that investment. In other words, the wind as a 'resource' did not exist before it was mapped and quantified—it had to be translated and inscribed and associated with certain values. To do this, the wind's 'value' or 'worth' must be stabilised and black-boxed by continuous framing/valuation processes. In and of itself the wind with its stochastically fluctuating speeds and shifting directions—something that has co-shaped local cultures and environments around the world—has no intrinsic value. Instead, this value must be constructed. That is, the wind must be calculated and attributed and associated/framed/valuated—and black-boxed—with a certain 'value' before it can become an extractable 'resource' 'out there' that can be harvested and exploited by some. This in turn requires that the investor becomes equipped with calculative tools—for example, EIA reports, Wind Atlas maps, atmospheric simulations, global circulation models, wind frequency distribution curves, wind turbine blades, power curves, software algorithms, price-setting mechanisms and financial spreadsheets full of key performance indicators (KPIs).

First, the path to becoming an energy resource starts with quantifying the *energy potential* of the wind. The first European Wind Atlas published by Risø National Laboratory in Denmark back in 1989 had a ground-breaking effect in that it helped to attract wind farm developers and others to areas where otherwise the wind resource was unknown. It thus helped to attract and mobilise a network, or a 'socio-technical (market) assemblage', around it (Caliskan & Callon, 2010; Labussiere & Nadai, 2018b; Kirkegaard et al., 2020; Kirkegaard, 2019). Based on a broad mapping of wind speeds at 50m height and with a knowledge of local terrain characteristics, it was possible to estimate the prospect of energy production and, combined with the existing electricity pricing scheme, potential revenues could be assessed. With a Wind Atlas, different sites around the world could be "qualitatively specified as an energy deposit, and their energy potential most often quantified" (Nadaï & Labussiere, 2018b, p. 57). Before deciding to invest in a new project, a wind farm developer would select promising candidate sites from a portfolio of sites based on Wind Atlas data in order to be able to estimate energy production and thus profits. The wind atlas analysis and application program, the software known as WAsP, helps to account for topography differences, landscape roughnesses and the built environment, allowing developers to design their wind farms and is today a whole suite of software packages.

Next, in order to be transformed from a potential energy resource to a commodity, turbine technologies help to make the wind extractable, calculable and tradable. The most obvious components are the very blades that extract the kinetic energy out of the wind and, by rotating the main generator shaft, convert it from mechanical to electrical energy and following the path of transforming the wind into a resource that can be harvested. Finally, once the energy has left the turbine then the path takes us to the electricity meter, where the flowing electrons are converted into a measurable quantity of energy: the kilowatt-hour. Once this transformation has taken place, specific pricing mechanisms such as feed-in tariffs (FIT)—so-called capitalisation instruments (Cointe & Nadaï, 2018)—help to commodify and marketise the wind through economisation processes, with investors guaranteed a fixed price (tariff) which was above the electricity market price (Caliskan & Callon, 2010; Cointe & Nadaï, 2018; Kirkegaard et al., 2020). In Denmark, the FIT encouraged the participation of a diversified group of actors, such as small stakeholders along with larger, more commercially orientated RET developers. After pressure from the European Union, more market-based pricing mechanisms were introduced, culminating in 2018 with the introduction of a tendering scheme of competitive bidding where developers compete to win the auctioned capacity onshore, on the basis of the lowest price per unit of energy produced (kilowatt-hour). While pushing down electricity prices, the tendering scheme favours large-scale corporate developers over smaller (community-based) actors in its selection criteria. The different pricing schemes provide data to wind farm developers that can be put into a spreadsheet and presented to a financial investor as KPIs of net present value (NPV), internal rate of return (IRR) and levelised cost of energy (LCoE). Here, the wind has been transformed into an assetised commodity that can be traded and marketised, with certain 'facts-based' or 'black-boxed' values attached, making it possible to form a socio-technical market assemblage around it.

The making of wind power resources does not come uncontested, however. The construction of values and frames involves the inclusion and exclusion of some values and actors, as evidenced in the Danish case above. That reconfiguration of a market assemblage around wind power in Denmark has produced contestation over the privatised commodification of wind power development (Kirkegaard et al., 2020). The matters of concern that the exclusion processes produce are in turn linked to underlying contestations over who should harness and 'own' the wind. This controversy is linked to what is known as the 'tragedy of the commons' (Ostrom, 2000), denoting the tendency of privatising and commodifying 'common resources' by 'enclosing' them with property rights and other means (ibid.). Issues of ownership and privatisation of commodities have also been dealt with to a great extent, and in great detail, in the SA literature, for example, by van der Horst and Vermeylen (2010, 2011, 2012). For instance, Van der Horst & Vermeylen have helped to shed light on energy-landscape conflicts by focusing on discrepancies between private and social benefits of maximised energy capture (2010), for example, adopting an analytical property rights framework that cuts across various levels of claims and value statements (2012). We argue that the ANT lens can complement the existing SA literature on this issue, which has been founded mostly in an institutional and/or social constructivist lens of narratives. With the inclusion of seemingly mundane material actors such as calculative devices into its analysis, displaying their constructive (performative) power, the ANT lens enables the analysis to open the black box of the wind's contested valuation, its economisation and marketisation, as well as its politicisation (Callon, 2007). Indeed, valuation devices and capitalisation instruments are not neutral props in the making of values (Dussauge et al., 2015, p. 268; Cointe & Nadaï, 2018), but highly political, causing controversy (Kirkegaard et al., 2020; Kirkegaard & Nyborg, 2020).

Snapshot 3: The Making of Stakeholders and Stakes

Lastly, ANT can help us to shed light on how 'stakeholders' and 'stakes' are black-boxed and framed by different authorities and their calculative devices. The ANT lens makes it possible to explore how this work of black boxing is done through various market devices that are bound up with 'stakemaking' (Dussauge et al., 2015), that is, with the making of values and stating of 'what matters'.

As an example of (monetary and non-monetary) stakemaking, we can take a look at the municipal planning system and the use of devices, tools and models to frame and value stakes and stakeholders. A powerful example is the (normative) map of 'The good process' in the Danish planning system (Danish Wind Industry Association, Local Government Denmark, and The Danish Society for Nature Conservation, n.d.), which depicts a seemingly linear and non-conflictual model for how and when to engage specific 'stakeholders' such as local residents in the planning and development of a wind farm. The model delineates specific phases in the planning of a wind farm site during which local community actors can be engaged and participate in public consultation processes (during the idea phase, the EIA phase and the project development/ implementation phase). Meanwhile, outside these phases the local communities do not have a right to argue for 'their stake' (Clausen et al., 2021), and when submitting complaints during the consultation process, many of the concerns raised are deemed illegitimate as they do not fall within the assigned techno-regulatory or technical-rational closed space (Lefebvre, 1991). The EIA, entangled in a planning system which frames stakeholders as those within a given geographic limit, has the power to define not only 'who' is in and out, but also 'what' is at stake, that is, what types of concerns or issues that should be accounted for is determined by EIA experts. Stakeholders and their stakes are not pre-given, but socio-materially construed.

The way in which developers and municipalities have the power to decide which concerns are legitimate and which are not, also relates to the SA literature, which has offered rich critique of the notion of 'Not-in-my-Backyard' (NIMBY), showing how the EIA and planning process allow legitimising some concerns and not others (e.g. Bristow et al., 2012; Cotton & Devine-Wright, 2012; Devine-Wright, 2011). Here, stakeholders are, for example, construed as 'communities of place' versus 'communities of interest' (Devine-Wright, 2011). An ANT lens can complement the SA literature by shedding light on the role of material actors in 'stakemaking', such as the role of calculative valuation devices, be it through public engagement models ('the good process', for instance), planners' categorisation of people's concerns, EIA measurements, risk indexes and more. The ANT lens can help to open the black box of how values and concerns, and stakes, are constructed through the social-material apparatus of the markets (Kornberger et al., 2015), that is, through the calculative devices that help to frame some concerns within, and some outside, the wind power market assemblage, and how such processes of inclusion-exclusion can produce contestation. In response to contested 'stakemaking', local opposition throws open issues of stakes and value, that is, about 'what counts', in a market. While some engage in direct opposition, other stakeholders 'learn to be affected', 'playing the game' so their concerns fit into the techno-regulatory frame (Metzger, 2013, p. 781).[3] By using the ANT lens to trace the role of material and seemingly mundane tools of valuation that help to black box/frame stakeholders, we also see how stakes are produced; yet "What is supposed to be at stake, and what is at stake" "is the object of intense politics" (Dussauge et al., 2015, p. 273).

Conclusion: Emergence, Agency and Dynamics

Our three snapshots have served the purpose of showing and illustrating what an ANT lens can help to illuminate and thus complement the SA literature. By tracing the socio-material construction of taken for granted concepts such as wind power sites and landscapes, resources and stakeholders, the ANT lens helps to shed a critical light on the effect/outcome of black boxes and framings and their inherent valuations as well as on how they sometimes engender power struggles and 'non-acceptance'. In other words, the ANT lens has helped us to inquire into the *emergence, agency and dynamics of the black boxes.* We argue that it is necessary to inquire into *how* black-boxed constructs have been socio-materially constituted the way they have —essentialised as 'matters of fact'—as well as on the effects (performative power/agency) these constructs have, in order to understand how and why they engender controversy/social opposition to wind power that reconfigures and sometimes even transforms socio-technical assemblages around wind power (dynamics). Such valuation struggles and controversies may often be overlooked or mistaken for something else, as they tend to lie 'beneath the surface' if we do not have a lens that can capture it.

Firstly, an ANT lens can help to shed light on how seemingly natural wind energy sites have been 'created' through socio-material tools such as maps, calculations, facts, 'zoning' and landscape categorisations and so on that support (and perform) the making of landscapes/sites (issues of *emergence* and *agency*): these tools support certain agendas, strengthen certain networks and not others. In other words, wind energy landscapes are actively made. Furthermore, such socio-materially construed landscapes may trigger local opposition where local community members oppose the way in which landscape expertise and zoning practices marginalise their claims as illegitimate. Thus, the black-boxed 'fact' that certain areas are 'ideal for wind turbines' is contested, producing *dynamics* of overflowing.

Secondly, an ANT lens helps to critically inquire into the socio-material construction of wind energy resources. Following the employment of various calculative devices we showed how wind energy has been transformed from an untamed turbulent flow to a tameable resource and commodity (Labussiere et al., 2018, p. 247-8; Li, 2014; van der Horst & Vermeylen, 2010). This reflects processes of assetising and commodifying 'nature' (the wind) into a resource through the use of inscription devices (Li, 2014, p. 589; Labussiere et al., 2018, p. 243; Birch & Nadaï 2020) and the *emergence* of a socio-technical market assemblage around it. The wind energy resources for specific sites are thus assembled or 'made up' (Hacking 1986 in Li, 2014, p. 589), mobilised by heterogeneous actors with agency to valuate and to proclaim expertise and ownership (Li, 2014; Labussiere et al., 2018). Marketisation of 'common' resources through socio-material valuation practices can however be an underlying reason for public contestation over wind farm developments (*agency* and *dynamics*).

Thirdly, the ANT lens helped us to illustrate the making of stakeholders and stakes and their identities. The stakes (and values) of stakeholders in Denmark tend to be framed within the narrow technical-regulatory space (see Gaventa, 2006) of the public hearing process in the planning system (Clausen et al., 2021) by experts where only issues, values and concerns that can be targeted through technical, legal or economic modifications are deemed legitimate, whereas other concerns are largely overlooked, while being framed as illegitimate. Meanwhile, such black boxing of potential concerns tends to overflow into controversy. The story reveals how stakemaking is bound up with inherent politics (van Hoyweghen,

2013) and power struggles. By addressing stakemaking in practice, we can shed light on how matters of concern are constructed in terms of "conflicts about what the concern is", as well as in terms of "the correct way of assessing a stake" (Dussauge et al., 2015, p. 273), potentially destabilising the development of new wind farms *(emergence, agency, dynamics)*.

Contributions to Social Acceptance Literature—and a Call for Interdisciplinarity

Our insights from the three snapshots have only been possible by inquiring into the often overseen 'mundane' and material actors, such as calculative devices, which help to qualify, value and frame wind farm sites and landscapes, resources and stakeholders and their stakes. The ANT lens entails a lateral and symmetrical perspective whereby the analyst should take seriously the role of non-human actors. As its interest also includes non-human actors, the ANT lens helps to shed light on the socio-*materially* entangled configuration of wind power. By tracing the non-human actors and their performative work, the ANT analyst is able to demonstrate how and by what and whose interests are considered, which perspectives are taken into account, and how power and agency are distributed. ANT thus helps to offer a critical lens of power—one that sees politics (negotiations, contestations, controversy) potentially everywhere (Latour, 2005b, p. 260; Caliskan & Callon, 2010, p. 12), even in wind farm zoning practices, a software algorithm, a wind atlas map, EIA visualisations and so on. Technology, science, markets, economy and politics are entangled (Caliskan & Callon, 2010; Callon, 2007).

With this chapter, we wish to call for interdisciplinarity between the SA literature and ANT. The SA literature has provided rich insights into multiple aspects of RET developments, and reasons for acceptance and opposition, often from a socio-psychological and/or spatial outlook. Moreover it has shed light on regulatory, procedural and distributional aspects of the planning and development process (e.g. Wüstenhagen et al., 2007; Wind2050; Wolsink, 2007; Aitken, 2010; Devine-Wright & Howes, 2010; Ellis & Ferraro, 2016; Clausen et al., 2021; Lennon & Scott, 2015; Walker & Baxter, 2017; Cass et al., 2010; Aitken, 2010). Along with Kirkegaard et al. (2020), we argue that the SA literature can be complemented by a lens of socio-technical assemblages. Thus, the key contribution from an ANT lens is obviously not in paying attention to the differing meanings or symbolic attributes of landscapes, locations, spaces, nature or the site that host wind turbine parks, in pointing out the distributional injustices inherent in exploiting wind as a resource in current regulative frameworks and markets, or in dissecting the way certain concerns are deemed illegitimate in the current planning system. In all these cases and analyses, all agency is granted to humans and the object of analysis is what humans do, make sense of, attribute meaning to, experience or construct. What an ANT lens does is to put material agency at the front and centre. It forces the analyst to acknowledge the (potential) agency of—and inherent black-boxed relations built into—hedgerows, ground water, birds, maps, zoning categories, Wind Atlas maps, atmospheric simulations, light flicker and so on in constructing and maintaining certain relationships and lending power to certain framings and categorisations. The ANT lens can, in other words, tell us something about social acceptance of wind power by making connections between a wind atlas map or a dike construction and a local siting conflict, which would otherwise have remained unexplored if sticking to a primary, human-centred, focus on 'community acceptance', market acceptance or socio-political acceptance and detangling these forms of acceptance from each other and from the material actors they co-exist with (Wüstenhagen et al., 2007; for a full critique, see Kirkegaard et al., 2020).

We acknowledge that there are ontological and epistemological discrepancies between ANT and most of the SA literature. Yet, like Hess and Sovacool (2020), we do see a benefit from cross-fertilising the two perspectives, boding well for further critical inquiries into processes of constructing, maintaining and transforming wind energy sites, landscapes, resources and stakeholders and their stakes, and much more.

Notes

1. Processes of site-making have also been examined from the viewpoint of the political and rhetorical dimensions of certain discourses, but combined with a socio-material lens of ANT (see, e.g., Rudolph & Kirkegaard, 2019).
2. Earlier versions of this research have been presented in Nyborg et al. (2016).
3. Based on a discursive, socio-constructionist approach, the SA literature has also been important for highlighting the strategic use of RET-related discourses and the performative/political/strategic dimension of place (Batel et al., 2015; Haggett & Futák-Campbell, 2011; Usher, 2013; van der Horst & Vermeylen, 2011).

References

Aitken, M. (2010). Why we still don't understand the social aspects of wind power: A critique of key assumptions within the literature. *Energy Policy, 38*(4), 1834–1841.

Aaen, S. B., Kerndrup, S., & Lyhne, I. (2016). Beyond public acceptance of energy infrastructure: How citizens make sense and form reactions by enacting networks of entities in infrastructure development. *Energy Policy, 96*, 576–586.

Batel, S., Devine-Wright, P., Wold, L., Egeland, H., Jacobsen, G., & Aas, O. (2015). The role of (de-)essentialisation within siting conflicts: An interdisciplinary approach. *The Journal of Environmental Psychology, 44*, 149–159.

Batel, S. (2018). A critical discussion of research on the social acceptance of RET generation and associated infrastructures and an agenda for the future. *Journal of Environmental Policy & Planning, 20*(3), 356–369.

Birch, K. & Nadaï, A. (2020). Turning Sunlit Rooftops and Windy Sites into Energy Assets. Chapter 6 in Birch, K. and Muniesa, F. (eds.) Assetization: Turning Things into Assets in Technoscientific Capitalism, MIT Press. https://doi.org/10.7551/mitpress/12075.003.0009

Brighenti, A. M. (2018). The Social Life of Measures: Conceptualizing Measure–Value Environments. *Theory, Culture & Society, 35*(1), 23–44.

Bristow, G. I., Cowell, R. J. W., & Munday, M. C. R. (2012). Windfalls for whom? The evolving notion of 'community'in community benefit provisions from wind farms. *Geoforum, 43*(6), 1108–1120.

Caliskan, K., & Callon, M. (2010). Economization, Part 2: A research programme for the study of markets. *Economy and Society, 39*(1), 1–32.

Callon, M. (Ed.). (1998). *Laws of the Markets.* Blackwell Publishers/The Sociological Review.

Callon, M. (1991): 'Techno-economic networks and irreversibility'. In Law, J. (ed.): A Sociology of Monsters: Essays on Power, Technology and Domination. London: Routledge: 132–161.

Callon, M. (2007). An Essay on the Growing Contribution of Economic Markets to the Proliferation of the Social. *Theory, Culture & Society, 24*(7–8), 139–163.

Callon, M., & Muniesa, F. (2005). Peripheral Vision: Economic Markets as Calculative Collective Devices. *Organization Studies, 26*(8), 1229–1250.

Cass, N., Walker, G., & Devine-Wright, P. (2010). Good neighbours, public relations and bribes: The politics and perceptions of community benefit provision in RET development in the UK. *Journal of Environmental Policy and Planning, 12*(3), 255–275.

Chilvers, J., & Longhurst, N. (2016). Participation in transition(s): Reconceiving public engagements in energy transitions as co-produced, emergent and diverse. *Journal of Environmental Policy & Planning, 18*(5), 585–607.

Chilvers, J., Palletta, H., & Hargreaves, T. (2018). Ecologies of participation in socio-technical change: The case of energy system transitions. *Energy Research and Social Science, 42*, 199–210.

Clausen, L. T., Nyborg, S., Rudolph, D. P. (2021). The good process or the great illusion? A spatial perspective on public participation in Danish municipal wind turbine planning. Submitted to *Journal of Environmental Policy and Planning*.

Cointe, N., & Nadaï, A. (2018). The Politics of Some Policy Instruments. In O. Labussiere & A. Nadaï (Eds.), *Energy Transitions. A Socio-Technical Inquiry* (1st ed., pp. 143–190). Palgrave Macmillan.

Cordula Kropp, (2017) Controversies around energy landscapes in third modernity. *Landscape Research 43*(4):562–573

Cotton, M. D., & Devine-Wright, P. (2012). Making electricity networks 'visible': industry actor representations of 'publics' and public engagement in infrastructure planning. *Public understanding of science, 21*(1), 17–35.

Cowell, R. (2010). Wind power, landscape and strategic, spatial planning—The construction of 'acceptable locations' in Wales. *Land Use Policy, 27*(2), 222–232.

Danish Wind Turbine Owners' Association, Local Government Denmark, and The Danish Society for Nature Conservation. (n.d.). Den gode proces, https://ipaper.ipapercms.dk/Windpower/Dengodeproces/#/.

Devine-Wright, P. (2011). From Backyards to Places: Public Engagement and the Emplacement of Renewable Energy Technologies. In P. Devine-Wright (Ed.), *Renewable Energy and the Public. From NIMBY to participation*. Routledge.

Devine-Wright, P., & Howes, Y. (2010). Disruption to place attachment and the protection of restorative environments: A wind energy case study. *Journal of Environmental Psychology, 30*(3), 271–280.

Dussauge, I., Helgesson, C. F., & Lee, F. (2015). Valuography. Studying the making of values. In C. F. Helgesson, I. Dussauge, & F. Lee (Eds.), *Value Practices in the Life Sciences and Medicine* (pp. 265–283). Oxford University Press.

Ellis, G., & Ferraro, G. (2016). The social acceptance of wind energy: Where we stand and the path ahead. *JRC Science for Policy Report, European Commission*.

Finch, J., Geiger, S., & Reid, E. (2017). Captured by technology? How material agency sustains interaction between regulators and industry actors. *Research Policy, 46*(1), 160–170.

Geiger, S., & Gross, N. (2018). Market Failures and Market Framings: Can a market be transformed from the inside? *Organization Studies, 39*(10), 1357–1376.

Geiger, S., Harrison, D., Kjellberg, H., & Mallard, A. (Eds.). (2014). *Concerned Markets: Economic Ordering for Multiple Values*. Edward Elgar Publishing.

Haggett, C., & Futák-Campbell, B. (2011). Tilting at windmills? Using discourse analysis to understand the attitude-behaviour gap in renewable energy conflicts, Механізм регулювання економіки, 1.

Hess, D. J., & Sovacool, B. K. (2020). Sociotechnical matters: Reviewing and integrating science and technology studies with energy social science. *Energy Research & Social Science, 65*, 1–17.

Jolivet, E., & Heiskanen, E. (2010). Blowing against the wind—An exploratory application of actor network theory to the analysis of local controversies and participation processes in wind energy. *Energy Policy, 38*, 6746–6754.

Kirkegaard, J. K. (2019). *Wind Power in China—Ambiguous Winds of Change in China's Energy Market.* Routledge.

Kirkegaard, J. K., Cronin, T., Nyborg, S., & Karnøe, P. (2020). Paradigm shift in Danish wind power—the (un)sustainable transformation of a sector. *Journal of Environmental Policy and Planning.*

Kirkegaard, J. K., & Nyborg, S. (2020). The (re)birth of a social movement—the Frankenstein story of Danish wind power development. Paper presented at the 36th EGOS Colloquium, *Organizing for a Sustainable Future: Responsibility, Renewal & Resistance*, University of Hamburg, July 2-4, 2020.

Kornberger, M., Justesen, L., Madsen, A. K., & Mouritsen, J. (Eds.). (2015). *Making Things Valuable.* Oxford University Press.

Krauss, W. (2010). The "Dingpolitik" of wind energy in Northern German landscapes: An ethnographic case study. *Landscape Research, 35*(2), 195–208.

Labussiere, O., & Nadai, A. (2018a). Spatialities of the energy transition: Intensive sites making earth matter? *Energy Research & Social Science, 36*, 120–128.

Labussiere, O., & Nadai, A. (Eds.). (2018b). *Energy Transitions. A Socio-Technical Inquiry.* Palgrave Macmillan.

Labussière, O., Banos, V., Fontaine, A., Verdeil, E., & Nadaï, A. (2018). The Spatialities of Energy Transition Processes. In O. Labussiere & A. Nadaï (Eds.), *Energy Transitions. A Socio-Technical Inquiry* (1st ed., pp. 239–275). Palgrave Macmillan.

Latour, B. (2004). Why Has Critique Run out of Steam? From Matters of Fact to Matters of Concern. *Critical Inquiry, 30*, 225–248.

Latour, B. (2005a). From Realpolitik to Dingpolitik or how to Make Things Public. In B. Latour & P. Wiebel (Eds.), *Making Things Public: Atmospheres of Democracy.* MIT Press.

Latour, B. (2005b). *Reassembling the Social: An Introduction to Actor-Network-Theory.* Oxford University Press.

Latour, B. (2017). *Facing Gaia. Eight Lectures on the New Climatic Regime.* Polity Press.

Lefebvre, H. (1991). *The Production of Space.* Blackwell Publishing.

Lennon, M., & Scott, M. (2015). Contending expertise: An interpretive approach to (re)conceiving wind power's 'planning problem'. *Journal of Environmental Policy & Planning, 17*(5), 593–616.

Metzger, J. (2013). Placing the stakes: the enactment of territorial stakeholders in planning processes. *Environment and Planning A, 45*, 781–796.

Nadaï, A., & Labussiere, O. (2010). Introduction: Landscapes of Energies. *Landscape Research, 35*(2), 143–155.

Nadaï, A., & Labussiere, O. (2018). New energy resources in the making. In O. Labussiere & A. Nadaï (Eds.), *Energy transitions: A socio-technical inquiry* (pp. 49–100). Palgrave Macmillan.

Nadaï, A., & van der Horst, D. (2010). Introduction: Landscapes of Energies. *Landscape Research, 35*(2), 143–155.

Nyborg, S., Kirkegaard, J. K., Clausen, L. T., Jørgensen, M. S., & Rudolph, D. P. (2016) Social contention in Denmark over alternative wind power development paths. Paper presented at the 4S/EASST Conference 2016: *Science and Technology by Other Means*, Barcelona, August 31—September 3, 2016.

Ostrom, Elinor (1990). Governing the Commons: The Evolution of Institutions for Collective Action. Cambridge, UK: Cambridge University Press. ISBN 978-0-521-40599-7.

Roscoe, P., & Townley, B. (2016). Unsettling issues: valuing public goods and the production of matters of concern. *Journal of Cultural Economy, 9*(2), 121–126.

Rudolph, D. P., & Kirkegaard, J. K. (2019). Making space for wind farms: Practices of territorial stigmatisation in rural Denmark. *Antipode, 51*(2), 642–663.

Tania Murray Li, (2014) What is land? Assembling a resource for global investment. *Transactions of the Institute of British Geographers 39*(4):589–602

Usher, M. (2013). Defending and transcending local identity through environmental discourse. *Environmental Politics, 22*(5), 811–831.

van der Horst, D., & Vermeylen, S. (2010). Wind Theft, Spatial Planning and International Relations. *Renewable Energy Law and Policy Review, 1*(1), 67–75.

van der Horst, D., & Vermeylen, S. (2011). Local Rights to Landscape in the Global Moral Economy of Carbon. *Landscape Research, 36*(4), 455–470.

van der Horst, D., & Vermeylen, S. (2012). Ownership claims, valuation practices, and the unpacking of energy-landscape conflicts. *International Review of Sociology, 22*(3), 429–445.

van Hoyweghen, I. (2013). On the Politics of Calculative Devices: Performing life insurance markets. *Journal of Cultural Economy, 7*(3), 334–352.

Walker, C., & Baxter, J. (2017). It's easy to throw rocks at a corporation': Wind energy development and distributive justice in Canada. *Journal of Environmental Policy & Planning, 19*(6), 754–768.

Walker, G., Burningham, K., & Barnett, J. (2010). Renewable energy and sociotechnical change: Imagined subjectivities of 'the public' and their implications. *Environment and Planning A, 4*, 931–947.

Wind2050, http://www.wind2050.dk/

Wolsink, M. (2007). Wind power implementation: The nature of public attitudes; Equity and fairness instead of 'backyard motives'. *Renewable and Sustainable Energy Reviews, 116*, 1188–1207.

Woods, M. (2003). Conflicting Environmental Visions of the Rural: Windfarm Development in Mid Wales. *Sociologia Ruralis, 43*(3), 271–288.

Wüstenhagen, R., Wolsink, M., & Bürer, M. J. (2007). Social acceptance of RET innovation: An introduction to the concept. *Energy Policy, 35*, 2683–2691.

CHAPTER 7

Social Acceptance and Interdisciplinarity: Understanding the Constructive Power of Terminology

Claire Haggett

Introduction

It is certainly the case that no one discipline can encompass a full understanding of people's responses to energy infrastructures, and that working across disciplines can be of great benefit. However, there are some key issues with interdisciplinary working. This chapter focuses on one of them: terminology use across disciplines.

This chapter will demonstrate that the same terms—such as 'attitudes' and 'behaviour'—are used differently across a range of disciplines; and crucially, that this matters. Terminology frames how issues should be appropriately investigated; and can actually shape the very nature of the research problem itself. This means that the way in which different disciplines work, and the concepts and terminology they use, can construct the 'problem', as well as the approach to understanding it.

C. Haggett (✉)
The University of Edinburgh, Edinburgh, UK
e-mail: claire.haggett@ed.ac.uk

S. Batel, D. Rudolph (eds.), *A critical approach to the social acceptance of renewable energy infrastructures*,
https://doi.org/10.1007/978-3-030-73699-6_7

For example, much work on social responses to renewable energy technologies (RET) has considered the 'social gap' (following Bell et al., 2005), whereby a mismatch is purported to exist between public support for RET and low siting success. Possible explanations depend on understandings of attitudes and behaviours; and there can only be a gap between attitudes and behaviour if these are defined in very specific ways: that an attitude is a declared intention to behave in a certain way, and behaviour is a measurement of that intention being carried out. Similarly, there can only be a social gap if surveys of general attitudes are a good way to understand responses (Bell et al., 2013). This chapter is therefore an attempt to explore the terms used by various disciplines, drawing out the differences, and highlighting possible ways of productively working in the future.

Disciplines and Terminology

Disciplines emerge with different fields of study, concepts, methods—and terminology. Often the same terms can refer to different ideas. This section considers some of the key terms used in research on social responses to RET—attitudes, and behaviours—from sociology, psychology, geography, and economics; and then selects examples of work using those concepts from each discipline. It then considers the implications of those differences.

Attitudes and Behaviours in Sociology

In sociology, attitudes may be prescriptions for action, can incorporate elements of learning, and can be indicative of an underlying value or belief about an object (Marshall, 1999; Jary & Jary, 1991, p. 32). Behaviour is not necessarily driven by conscious deliberation or 'choice', or preceded by intention, but is the product of habit, 'routines' or practices that structure society (Upham et al., 2015). Indeed, the focus in sociology is on the social context, and the social, cultural, political, and technological influences on attitudes and behaviours. Johnson (1995, p. 18) outlines that "as a concept, attitude is important because it incorporates the role of emotion and the power of social systems to shape, regulate and evoke it, producing both social cohesion and conflict".

Across sociological theory, there are differences in the conceptualisation of attitudes. Quantitative approaches may view attitudes as a guide to more deeply held values or to actual behaviour; in qualitative work such measurements may be considered relatively superficial. Expressed attitudes may be inconsistent with subsequent behaviour; and a specific measurable concept of an 'attitude' lacking in validity, with more reliable data elicited by more prolonged contact with respondents (Abercrombie et al., 2000, p. 19; also Marshall, 1999).

Sociological Work in Practice

These different understandings are reflected in sociological work on RETs. Some work assumes that attitudes are specific and measurable concepts; and their relation to each other can be hypothesised and tested. For example, Bidwell (2013) explores the relationship between values and beliefs in shaping attitudes towards the potential wind farms; measuring attitudes by asking participants on a five-point scale about different aspects of a proposed development. The study finds a gap between underlying values, where "altruistic values have a buoying effect on wind energy attitudes, while values of traditionalism diminish wind energy support" (2013, p. 198); and this gap exists because attitudes are conceptualised as being separate from values.

Other research places attitudes in the social context from which they arise. For example, Wheeler describes how attitudes towards wind farms are contextualised "within participants' wider relationships with the rural place in which they live" (2017, p. 110; see also Waldo, 2012, on the differences between attitudes, values, and behaviours in sociological research). However, other sociological research views attitudes quite differently. Shove (2010) argues that assuming that attitudes drive behaviour leading to choices misses the way in which context fundamentally shapes all of these, and that an approach which focuses on social practices is far more useful, considering energy as part of a complex system of social practices (Shove & Walker, 2014). A focus on attitudes would therefore mean ignoring questions about what energy means or the sets of social practices on which energy demand depends.

Attitudes and Behaviours in Psychology

While the concept of 'attitudes' is applied loosely and differently across sociology, it is used extensively and technically in psychology (Abercrombie et al., 2000). Individuals respond to various 'attitudinal objects' (other individuals, social groups, situations, social issues), and this response is predetermined by their attitude towards that particular object (Howarth, 2006, pp. 693–694). Attitudes typically comprise three components: cognition (knowledge and beliefs), affect (emotional response), and behaviour (past and current behavioural response) (Upham et al., 2015); and are seen as relatively fixed and stable over time and across contexts (Hogg & Vaughan, 2002).

Attitudes are widely assumed to influence behaviour (Eyseneck, 1994, p. 30), and to exert a dynamic and directive influence (Pettijohn, 1991, p. 22). Any given attitude-relevant behaviour is influenced by a multitude of additional factors. Research has explored the conditions which provide correlations between global attitudes and specific actions, and the nature of a particular attitude and the way in which it is acquired (Kupar & Kupar, 1996). The best-known model of this is the Theory of Reasoned Action (Fishbein & Ajzen, 1975), which correlates specific behaviours with specific attitudes, as well as taking account of social norms. So, the global attitude of feelings about wind power would not be used to predict behaviour towards a proposal; the attitude about having a wind farm nearby would be used instead. In this model, individuals carefully appraise the information they have about any behaviour, form attitudes and beliefs on the basis of this information, and then act in ways that are consistent with their considerations. An extension of this is the Theory of Planned Behaviour (Ajzen, 1988), which acknowledges that intentions depend on individuals having sufficient control over the behaviour in question.

Howarth (2006) also describes differences within psychology, with an 'Americanization of the attitude concept' tied to discourses of individualism and subjective evaluation, compared to a social representations approach which focuses on community, collective practices and the institutionalisation of social knowledge. She notes that the individualisation of social psychology has led to an extremely narrow understanding of attitudes focused almost entirely on the decontextualised individual, and hence an (almost) asocial and apolitical version of social beings. This, in turn, has led to the development of a variety of theories attempting to 'put the social back in social psychology', with a broadening the concept of attitudes.

A further difference exists with discursive psychology. Edwards and Potter (1992) outline that a 'discursive action model' means a radical reworking of some of psychology's most central concepts such as language, cognition, truth, knowledge, and reality. Potter and Wetherell (1987, p. 49) have argued that "discourse analysis does not take for granted that accounts reflect underlying attitudes or dispositions". This approach does not attempt to try and reveal inner mental states or attitudes (or even make assumptions about their existence), and instead focuses on how such things are presented and referred to in accounts.

Psychological Research in Practice

There has been a wealth of psychological research on social responses to RET. For example, Jones and Eiser (2009) used multiple regression analysis to identify factors that predict attitudes towards proposed RETs. Walter (2014) explored how intentions to act are based on attitudes towards RET, finding that general attitudes towards wind energy influence local acceptance of a particular project. Bang et al. (2000) draw on the theory of reasoned action to explain concern, knowledge, belief, and attitudes towards renewable energy. They devise three variables—environmental concern, RET knowledge, and beliefs about the consequences of RET—and correlate these with attitudes towards paying more for renewable energy, and find relationships between beliefs and attitudes. Huijts, Molin and Steg (2012, pp. 525–566) developed a model of psychological factors that influence attitudes (acceptability) and behaviours (acceptance) for or against technologies. They suggest that intention to act is influenced by attitudes, social norms, perceived behavioural control, and personal norms; while attitudes are influenced by the perceived costs, risks and benefits, positive and negative feelings in response to the technology, trust, procedural fairness and distributive fairness.

Research not in disciplinary journals but in topic-based publications (such as *Energy Policy*, or *Wind Energy*) also draws on psychological ideas about attitudes and behaviours. For example, Johansson and Laike's (2007) study uses psychological theories to set out that attitudes are measurable concepts, they form the basis of behaviour, and that appropriate actions can be undertaken on the basis of analysis of them.

Attitudes and Behaviours in Geography

There is an emphasis on the relatively stable and consistent nature of attitudes in geography. An attitude may be understood as "a state of mind, and a relatively enduring tendency to perceive, feel or behave towards certain people or events in a particular manner". Attitudes play an "important part in perception and they affect preferences and the choice of goals"

(Clark, 1995, p. 39), and they often imply that a value judgement has been made (Goodall, 1997).

Geography does not necessarily make a distinction between action and behaviour; and the subdiscipline of behavioural geography focuses on the way that people's perceptions of their environment may not correspond with 'objective reality' (Mayhew & Penny, 1992). Goodall describes that behaviour is not a simple and straightforward reaction to the 'objective' environment (where modifications are made to the natural environment by humans) but a reaction to a "partial and distorted psychological representation of that environment" (1997, p. 37).

Geography is a broad discipline, and in response to behavioural geography, a 'practice-turn' has emerged, which emphasises the wider social context which shapes responses. As Ellis and Ferraro (2016, p. 3) say, "methods used in many attitude studies have tended to constrain the understanding of the social, dynamic and geographic complexity that shapes acceptance", and efforts are therefore being made to redress this.

Geographical Work in Practice

There has been extensive work in geography on responses to RETs. For example, Warren et al.'s (2005) classic 'green on green' paper conceptualised clashing local and global environmental concerns, drawing on case studies about 'public attitudes' to wind farms. It explored influences upon attitudes, such as perceptions of visual impact, and inadequate planning measures. Van der Horst (2007) discusses the issues that influence attitudes to proposed projects, and argues that individual and group attitudes are different concepts, determining the factors that 'on aggregate' affect 'public attitudes'. Friedl and Reichl (2016) explore the attitudes, perspectives and positions of key stakeholders, and the factors which shape positive and negative attitudes towards energy projects. They focus on attitudes in a broad context and conduct qualitative analysis to gain a holistic understanding of social acceptance.

Attitudes and Behaviours in Economics

There is no entry for 'attitude' in economics dictionaries and encyclopaedias. Disagreement seems to exist about whether attitudes are a measurable concept or one of many factors that feed into preferences; but behaviour is a key focus. In neoclassical economics, assumptions are made that individuals seek to maximise their utility, drawing on perfect information, and make trade-offs between various attributes to form preferences. Behavioural economics brings in psychological elements to explore ideas around economic rationality, often stressing the 'irrational' aspects of decision making. Research includes 'choice experiments' and exploring individuals' 'willingness to pay' for a range of alternatives (Eatwell et al., 1995; Pollitt & Shaorshadze, 2011).

There is a range of economic work on responses to RETs, which sets out a difference between public and individual attitudes towards projects, and that influences on these can be determined (e.g., Ek, 2005; Ladenburg, 2010). For example, Ek and Persson (2014) developed a choice experiment asking individuals to choose between two hypothetical wind farms characterised by different attributes. Ladenburg's (2008) respondents were asked to express their attitude towards more wind farms from very positive to very negative. In later work, models are applied to results to determine the significance and frequency of responses, and results find that attitudes are formed by a variety of demographic, locational, and experiential factors (such as frequency and type of visit to areas with wind farms) (Ladenburg, 2010).

Overview of Some Disciplinary Differences

There are some common elements in the conceptualisations of attitudes and the relationship to behaviours. There is some agreement that attitudes include an evaluative element and an element of learning; but—there are very significant differences within and between disciplines. There are differences about the emergence, existence, and strength of the relationship between attitudes and behaviours; the role of social context and wider influences; the stability of attitudes; and whether global or specific attitudes are most important (or if these are even meaningful concepts).

Importantly, there is also considerable disagreement about if and how to measure attitudes. Discursive psychology does not attempt to determine attitudes, and believes that attempts to do so structure and constrain the very thing they are attempting to understand. Qualitative sociologists would argue that attitudes can only be understood by in-depth research as part of a broader set of concepts, while quantitative sociologists and work in psychology has developed increasingly sophisticated scales and measures. Some of these differences reflect the broader ontological differences between qualitative and quantitative work in general, as well as differences in disciplines (Upham et al., 2015).

Perhaps even more fundamentally, there is also considerable disagreement about what the concept of an 'attitude' means. The same term—'attitude'—is used to mean different things: a specific measurable entity which correlates to behaviour; or a vaguer concept that might only loosely be connected; or even a meaningless label which obscures a focus on practices. Significantly, the term can be used as a substitute for the concepts of 'beliefs' and 'values', or an expression of emotion. As will be discussed, this is not merely a matter of semantics but something much more fundamental to the research being conducted.

The Application of Concepts and Terminology in Research

It is perhaps unsurprising that disciplines understand and use terms and concepts differently, as this is intrinsically part of the distinction between them. What is important—and perhaps more surprising—is that these differences are often ***not defined*** in research.

As set out above, the terms 'attitude' and 'behaviour' can mean very different things in different disciplines; and these terms are extremely well used throughout the now extensive literature on social responses to energy infrastructures. The term 'attitude' pervades. Attitudes are commonly documented, described, and researched in a range of work on social acceptance of renewables, in research published in a range of different topic-based and disciplinary journals. Very often this seems to be utilising a common-sense understanding of what this might mean; attitudes are merely 'what people think'; at other times, this is very specifically used.

What is significant is that, even when focused *explicitly* on attitudes and behaviour—some research does not define what is meant by 'attitude' or 'behaviour'. For example, Owens and Driffill (2008) set out that they "critically reflect on the way it is often assumed that 'attitudes' and 'behaviours' need to change; and that there exist pre-determined goals around which modified attitudes and behaviours should be shaped". They discuss the way in which a focus on attitudes and behaviours misses the complexity of decision making and the "social, cultural and institutional contexts in which attitudes and behaviours are formed" (2008, p. 4413)—but—without ever once setting out what is meant by an attitude, or a behaviour.

Similarly, Warren and McFadyen's (2010) study of the effect of community ownership of wind farms on "public attitudes" to wind energy covers their understanding and use of the term 'local acceptance' in some detail, but not once what they mean by 'attitude'. Kaldellis et al.'s (2012) paper assesses 'the public attitude' using a series of questionnaires. The title of the research is "What is the public attitude?" but the term is not defined, and seems to be synonymous with 'opinion' or 'what people think'—and is not made explicit. Jones and Eiser's (2009) psychological research on RET uses 'attitude' and 'opinion' interchangeably; Eltham et al. (2008) explore '*changes* in attitudes' to a wind farm; use the term 'opinion' synonymously, and state that both correlate with acceptance—with none of these terms explained. Upham, Oltra and Boso (2015, p. 103) seem to conflate terms together when they base their definition of acceptance on "a favourable or positive response (including attitude, intention, behaviour and—where appropriate—use) relating to a proposed or in situ technology or socio-technical system, by members in a given social unit".

This widespread lack of definition—and perhaps also lack of precision—seems remarkable. Studies specifically set out to explore attitudes and behaviours and yet do not state what is meant by these terms. These uses of terminology matter for understanding the results and resonance of the research, and for building on research—for how can this reliably be done, if it is not entirely clear what has been assessed?

The Importance of Conflicted and Conflated Meanings

As outlined above, there are significant differences within research on this topic; between conceiving of attitudes and behaviours as concepts that can be understood, and examining the correlation between them; or using commonplace meanings of these ideas, for example, taking attitude as a synonym for opinion or value.

These differences are not often articulated—and this is important because it makes research findings hard to compare. As Castro says, "different studies use contrasting definitions for the same concepts. As a result, the epistemological assumptions of the studies are not always clear and different authors frequently assume that the same instruments assess different concepts" (2006, p. 251). Indeed, Castro points out that in the research literature on environmental concern, contrasting definitions and measurements for the same concepts prevented an advancement in understanding. This surely has to also apply to this topic area.

For example, one of the key points from psychology is that attitudes and behaviour have to be measured and compared at the same level. A global attitude about wind farms may not relate to a specific behaviour. But how can it be known whether on what level an attitude is being measured if this is not set out? Similarly, if the term 'attitude' can also act as a substitute for 'beliefs' and 'values', how can it be known what is being measured; and how can it be compared to other research?

This lack of clarity, and conflation of differences, matters because the disciplinary principles or methodological approaches are often obscured or ignored when summaries of research findings are presented, or the state of the field is being précised. 'What do we know now?' is presented as the findings, the results, the knowledge—not the basis on which it was generated. So, the range of factors that might influence social responses becomes the headline; and becomes the points that are used in subsequent literature summaries to explain and justify a new project. But, of course, the results generated depend on the methods used; and without knowing more about that basis, it actually becomes very hard to know how to interpret the headline results, and therefore to build on the findings. Upham et al. (2015) rightly state that interpreting results obtained via diverse methods and perspectives raises questions about complementarity and integration; surely even more so if the terms used for those results differ or are obscured.

The importance of this is seems sometimes missed; for example, when Bidwell (2013, p. 190) outlines that "scholars have urged a more nuanced understanding of public attitudes and motivations regarding the development of wind energy and other renewables"; what he means is more substantive research to identity the factors that influence attitudes, when perhaps what is (also) needed is a more nuanced understanding of what is meant by 'attitude' or greater clarity across approaches.

The Constructive Power of Terminology

The terms used in research matter not only because clarity is needed to be able to understand, and to compare, research findings. Terminology matters because it can shape the nature of the research project; and indeed, of the very phenomena that is being measured. If attitudes are perceived as measurable, and that they are determinants of behaviour, then examining the influences on them becomes the issue. If attitudes are synonymous with opinions, responses and behaviours, then a research project will look very different.

There has been valuable work on the role of terminology in research on RETs, and two papers are of particular significance here. Firstly, Burningham's (2000) seminal paper rightly examined the value-laden label 'NIMBY' (Not in my back yard), and how it actively constructed the role of opponents, presenting them as having the selfish parochial motives that the term implies.

Secondly, Batel et al. (2013) rightly highlight that the using word 'acceptance' is not neutral or unproblematic. This is absolutely correct; 'acceptance' has pejorative implications, and problematizes those who oppose. Moreover, Batel et al. (2013) show that the terminology of 'acceptance' shapes the scope and framing of the research; if understanding 'social acceptance' is the aim, then the questions ask respondents whether they accept or oppose a project. Batel, Devine-Wright and Tangeland go on to argue that research should therefore "critically reflect on its use of language to define the research problem" (2013, p. 2). This applies to all terminology. All language choices shape and frame the concept they purport to describe (see e.g., Billig, 1996; Wetherell, 1998); and this applies to the terminology used in research on RET.

For example, if considering 'the gap' between attitudes and behaviour, it is clear that whether there is a gap or not depends on the terms used for these concepts. There can only be a gap between an individual's attitudes and behaviour if these are defined in very specific ways; that an attitude is a declared intention to behave in a certain way, and behaviour is a measurement of that intention being carried out. The premise for Swofford and Slattery's (2010) paper is that previous research has found 'general' attitudes to RET are positive; and that this can be correlated to acceptance. But highlighting a difference which needs to be explained between individual and general attitudes is an artefact of understanding attitudes in this way. And a gap between attitudes and planning success disappears if national attitude surveys are not considered meaningful measures.

In addition, as discussed earlier, Bidwell's (2013) study finds a gap between values and attitudes, where "altruistic values have a buoying effect on wind energy attitudes, while values of traditionalism diminish wind energy support" (Bidwell, 2013, p. 198). But does this gap exist because of the way in which these concepts have been defined? If the concept of an 'attitude' just means 'what someone thinks', or could also include intentions and behaviours, then this precise gap between values disappears. Similarly, Bang et al. (2000) find relationships between beliefs and attitudes, because of the way in which they have defined those beliefs and attitudes; they ask questions to detect the extent of particular attitudes and explore how those feed into certain behaviours (or not). This means that way in which these concepts are understood actually shapes the research problem.

Implications for Future Work

What then does all of this mean, for understanding responses to RET? It is apparent that there are significant differences in terminologies, both in terms of different meanings of the same word, and different concepts (both within and between disciplines). What implications does this have for these disciplines working together?

Firstly, this may depend on what is intended. Areas of common ground between disciplines have been highlighted, for example, there is *some* agreement that attitudes exist, they are evaluative, they influence behaviour, and that values underlie this process. But this agreement is not predominant. Disciplines are different precisely because they have fundamentally different bases and principles. Sociology and psychology are in many ways diametrically opposed (as are of course qualitative and quantitative approaches). Bang et al.'s (2000) method, which correlated three particular variables with attitudes, simply would not be meaningful to a qualitative sociologist, who would explore the context in which responses emerge, rather than isolating particular ideas. Further, discursive psychology is fundamentally different from psychology, and developed in direct opposition to it. Setting out key concepts therefore may be difficult.

But perhaps there is some potential for complimentary work that builds on the different approaches that different disciplines can bring; a focus on social context then illuminated by detailed discourse analysis for example. As the premise for this Book rightly sets out, no single disciplinary approach is sufficient for understanding people's responses to RET, and combining the strengths that each brings may be a fruitful approach; as Batel and Castro (2018) interestingly suggest; and applied in practice by Haggett and Toke (2006).

Secondly, it is useful to think about how this might be done. It is clear that discussing terminology matters; words are not neutral windows on the world. There are different understandings of what the same terms mean—and these need to be articulated if disciplines can work together. Definitions are important not just to ensure that representatives from different disciplines can talk to each other; measuring attitudes may mean doing something quite different—assessing cognitive processes, or the outcome of social processes.

It is therefore important to follow the calls from Aitken (2010), Batel et al. (2013), and Jefferson (2018) that a more critical conceptualisation of research agendas and underpinning assumptions is required—and because terminology is so fundamental, this needs to be from the very start. There are perhaps three aspects to this. The first is to start from a position of acknowledging the terminology matters, and needs to be considered. Secondly, it means actually articulating the intended meanings—including them and with precision, rather than just assuming that their meaning will be understood. Thirdly, it means keeping definitions close at hand. As Howarth (2006) notes, words that may have common-sense meanings, but actually incorporate substantial elements within them, and it seems imperative to remain mindful of this. Some interdisciplinary approaches may be too far removed from each other to be successful; but an upfront focus on terminology is surely needed for any attempt at productive collaboration at all.

Conclusion

This chapter has provided an overview of the different ways in which the terms 'attitudes' and 'behaviour' are used. It has outlined the way in which they hold different meanings across, and within, different disciplines; and highlighted a few examples of these approaches in practice. These terms are prevalent throughout a wide range of research on social responses to renewable energy infrastructure—in discipline and topic-based journals, and in research explicitly based in a discipline and other subject focused research. It matters therefore to understand what these terms mean, and how they are being used.

As has been demonstrated, there are both different interpretations of these terms and missing definitions. Indeed, the extent to which the differences in meanings are ignored, or the meanings are assumed, seems quite surprising. While 'attitudes' and 'behaviour' may seem to be straightforward terms, they have precise and different definitions in different disciplines. This is important. It matters to know what these terms mean and how they are being used. It matters to be able to understand existing research. It matters to be able to build on that research. And it matters to be able to conceive of productive interdisciplinary research.

Finally, it matters because only for the workings of any research project, but for the conceptualisation of the nature of the phenomena under study. Maintaining a critical approach to research on responses to renewable energy therefore has to mean finding and maintaining a critical eye on the language and terminology used.

References

Abercrombie, N., Hill, S., & Turner, B. S. (2000). *The Penguin dictionary of sociology*. Penguin.

Aitken, M. (2010). Wind power and community benefits: Challenges and opportunities. *Energy Policy, 38*(10), 6066–6075. https://doi.org/10.1016/j.enpol.2010.05.062

Ajzen, I. (1988). *Attitudes, personality, and behavior*. Dorsey Press.

Bang, H.-K., Ellinger, A. E., Hadjimarcou, J., & Traichal, P. A. (2000). Consumer concern, knowledge, belief, and attitude toward renewable energy: An application of the reasoned action theory. *Psychology and Marketing, 17*(6), 449–468. https://doi.org/10.1002/(SICI)1520-6793(200006)17:6<449::AID-MAR2>3.0.CO;2-8

Batel, S., & Castro, P. (2018). Reopening the dialogue between the theory of social representations and discursive psychology for examining the construction and transformation. *British Journal of Social Psychology, 57*(4), 732–753. https://doi.org/10.1111/bjso.12259

Batel, S., Devine-Wright, P., & Tangeland, T. (2013). Social acceptance of low carbon energy and associated infrastructures: A critical discussion. *Energy Policy, 58*, 1–5. https://doi.org/10.1016/j.enpol.2013.03.018

Bell, D., Gray, T., & Haggett, C. (2005). The 'social gap' in wind farm siting decisions: Explanations and policy responses. *Environmental Politics, 14*(4), 460–477. https://doi.org/10.1080/09644010500175833

Bell, D., Haggett, C., Gray, T., & Swaffield, J. (2013). Re-visiting the social gap: public opinions and relations of power in the local politics of wind energy. *Environmental Politics, 22*(1), 115–135. https://doi.org/10.1080/09644016.2013.755793

Bidwell, D. (2013). The role of values in public beliefs and attitudes towards commercial wind energy. *Energy Policy, 58*, 189–199. https://doi.org/10.1016/j.enpol.2013.03.010

Billig, M. (1996). *Arguing and thinking: A rhetorical approach to social psychology*. Cambridge University Press.

Burningham, K. (2000). Using the language of NIMBY: A topic for research, not an activity for researchers. *Local Environment, 5*(1), 55–67. https://doi.org/10.1080/135498300113264

Castro, P. (2006). Applying social psychology to the study of environmental concern and environmental worldviews: Contributions from the social representations approach. *Journal of Community and Applied Social Psychology, 16*, 247–266. https://doi.org/10.1002/casp.864

Clark, A. N. (1995). *Longman dictionary of geography—Human and physical*. Longman.

Eatwell, J., Milgate, M., & Newman, P. (Eds.). (1995). *The new Palgrave dictionary of economics*. Macmillan.

Edwards, D., & Potter, J. (1992). *Discursive psychology*. Sage.

Ek, K. (2005). Public and private attitudes towards "green" electricity: The case of Swedish wind power. *Energy Policy, 33*(13), 1677–1689. https://doi.org/10.1016/j.enpol.2007.09.01

Ek, K., & Persson, L. (2014). Wind farms: Where and how to place them? A choice experiment approach to measure consumer preferences for characteristics of wind farm establishments in Sweden. *Ecological Economics, 105*, 193–203. https://doi.org/10.1016/j.ecolecon.2014.06.001

Ellis, G., & Ferraro, G. (2016). *The social acceptance of wind energy: Where we stand and the path ahead*. Publications Office of the European Union. https://doi.org/10.2789/696070

Eltham, D. C., Harrison, G. P., & Allen, S. J. (2008). Change in public attitudes towards a Cornish wind farm: Implications for planning. *Energy Policy, 36*(1), 23–33. https://doi.org/10.1016/j.enpol.2007.09.010

Eyseneck, M. (Ed.). (1994). *The Blackwell dictionary of cognitive psychology*. Blackwell.

Fishbein, M., & Ajzen, I. (1975). *Belief, attitude, intention, and behaviour: An introduction to theory and research*. Addison-Wesley.

Friedl, C., & Reichl, J. (2016). Realizing energy infrastructure projects—A qualitative empirical analysis of local practices to address social acceptance. *Energy Policy, 89*, 184–193. https://doi.org/10.1016/j.enpol.2015.11.027

Goodall, B. (1997). *The Penguin dictionary of geography*. Penguin.

Haggett, C., & Toke, D. (2006). Crossing the great divide: Using multi-method analysis to understand opposition to wind farms. *Public Administration, 84*(1), 103–120. https://doi.org/10.1111/j.0033-3298.2006.00495.x

Hogg, M., & Vaughan, G. (2002). *Social psychology*. Prentice Hall.

Howarth, C. (2006). How social representations of attitudes have informed attitude theories: The consensual and the reified. *Theory and Psychology, 16*(5), 691–714. https://doi.org/10.1177/0959354306067443

Huijts, N. M. A., Molin, E. J. E., & Steg, L. (2012). Psychological factors influencing sustainable energy technology acceptance: A review-based comprehensive framework. *Renewable and Sustainable Energy Reviews, 16*, 525–531. https://doi.org/10.1016/j.rser.2011.08.018

Jary, D., & Jary, J. (1991). *Collins Dictionary of sociology*. Harper Collins.

Jefferson, M. (2018). Exaggeration and/or denial: Twin towers of industrial and commercial interests supposedly intended to accelerate energy transition in the United Kingdom. *Energy Research & Social Science, 43*, 8–15. https://doi.org/10.1016/j.erss.2018.05.008

Johansson, M., & Laike, T. (2007). Intention to respond to local wind turbines: The role of attitudes and visual perception. *Wind Energy, 10*, 435–451. https://doi.org/10.1002/we.232

Johnson, A. G. (1995). *The Blackwell dictionary of sociology: A user's guide to sociological language* (2nd ed.). Basil Blackwell.

Jones, C. R., & Eiser, R. J. (2009). Identifying predictors of attitudes towards local onshore wind development with reference to an English case study. *Energy Policy, 37*(11), 4604–4614. https://doi.org/10.1016/j.enpol.2009.06.015

Kaldellis, J. K., Kapsali, M., & Katsanou, E. (2012). Renewable energy applications in Greece—What is the public attitude? *Energy Policy, 42*, 37–48. https://doi.org/10.1016/j.enpol.2011.11.017

Kupar, A., & Kupar, J. (Eds.). (1996). *The social science Encyclopaedia* (2nd ed.). Routledge.

Ladenburg, J. (2008). Attitudes towards on-land and offshore wind power development in Denmark: Choice of development strategy. *Renewable Energy, 33*(1), 111–118. https://doi.org/10.1016/j.renene.2007.01.011

Ladenburg, J. (2010). Attitudes towards offshore wind farms—The role of beach visits on attitude and demographic and attitude relations. *Energy Policy, 38*(3), 1297–1304. https://doi.org/10.1016/j.enpol.2009.11.005

Marshall, G. (Ed.). (1999). *The concise Oxford dictionary of sociology*. Oxford University Press.

Owens, S., & Driffill, L. (2008). How to change attitudes and behaviours in the context of energy. *Energy Policy, 36*, 4412–4418. https://doi.org/10.1016/j.enpol.2008.09.031

Pettijohn, T. F. (1991). *The Encyclopaedic dictionary of psychology* (4th ed.). Dushkin.

Pollitt, M. G., & Shaorshadze, I. (2011). *The role of behavioural economics in energy and climate policy*. Cambridge Work Paper on Economics, 1–29. https://doi.org/10.4337/9780857933683.00035

Potter, J., & Wetherell, M. (1987). *Discourse and social psychology: Beyond attitudes and behaviour*. Sage.

Shove, E. (2010). Beyond the ABC: Climate change policy and theories of social change. *Environment and Planning A, 42*, 1273–1285. https://doi.org/10.1068/a42282

Shove, E., & Walker, G. (2014). What is energy for? Social practice and energy demand. *Theory, Culture and Society, 31*, 41–58. https://doi.org/10.1177/0263276414536746

Swofford, J., & Slattery, M. (2010). Public attitudes of wind energy in Texas: Local communities in close proximity to wind farms and their effect on decision-making. *Energy Policy, 38*(5), 2508–2519. https://doi.org/10.1016/j.enpol.2009.12.046

Upham, P., Oltra, C., & Boso, À. (2015). Social acceptance of energy technologies, infrastructures and applications: Towards a general cross-paradigmatic analytical framework. *Energy Research and Social Science, 8*, 100–112. https://doi.org/10.1016/j.erss.2015.05.003

Van der Horst, D. (2007). NIMBY or not? Exploring the relevance of location and the politics of voiced opinions in renewable energy siting controversies. *Energy Policy, 35*, 2705–2714. https://doi.org/10.1016/j.enpol.2006.12.012

Waldo, A. (2012). Offshore wind power in Sweden—A qualitative analysis of attitudes with particular focus on opponents. *Energy Policy, 41*, 692–702. https://doi.org/10.1016/j.enpol.2011.11.033

Walter, G. (2014). Determining the local acceptance of wind energy projects in Switzerland: The importance of general attitudes and project characteristics. *Energy Research & Social Science, 4*, 78–88. https://doi.org/10.1016/j.erss.2014.09.003

Warren, C. R., Lumsden, C., O'Dowd, S., & Birnie, R. V. (2005). 'Green On Green': Public perceptions of wind power in Scotland and Ireland. *Journal of Environmental Planning and Management, 48*(6), 853–875. https://doi.org/10.1080/09640560500294376

Warren, C. R., & McFadyen, M. (2010). Does community ownership affect public attitudes to wind energy? A case study from south-west Scotland. *Land Use Policy, 27*(2), 204–213. https://doi.org/10.1016/j.landusepol.2008.12.010

Wetherell, M. (1998). Positioning and interpretative repertoires: Conversation analysis and post-structuralism in dialogue. *Discourse and Society, 9*, 387–412. https://doi.org/10.1177/0957926598009003005

Wheeler, R. (2017). Reconciling windfarms with rural place identity: Exploring residents' attitudes to existing sites. *Sociologia Ruralis, 57*(1), 110–132. https://doi.org/10.1111/soru.12121

PART V

Interventions—Praxis and Political Engagement with Research

CHAPTER 8

Social Acceptance: Beyond Criticism and Critical, a Call for Experimental Ontology

Alain Nadaï and Olivier Labussière

Introduction

Research on the social acceptance of renewable energy technologies (RET) started in the nineties, with successive 'waves' of research problematizing and critiquing the notion (Batel, 2020). While early works uncritically examined this issue with a view to promote acceptance—most often under the heading of the NIMBY notion—social sciences have then challenged the relevance of NIMBY. They argued that the notion lacks consideration for the potential of the 'social' to evolve, because it locates the potential for change in the technologies and the potential for resistance in the

A. Nadaï (✉)
International Research Center on Environment and Development (CIRED-CNRS), Paris, France
e-mail: nadai@centre-cired.fr

O. Labussière
CNRS, Pacte, Université Grenoble Alpes, Grenoble, France
e-mail: olivier.labussiere@umrpacte.fr

S. Batel, D. Rudolph (eds.), *A critical approach to the social acceptance of renewable energy infrastructures*,
https://doi.org/10.1007/978-3-030-73699-6_8

'social' (Batel et al., 2013; Shove, 2010). As a matter of fact, all developments in RET technologies confirm this criticism: new collectives and new relations emerge with these developments and witness of the potential of the 'social' to accompany, if not to take an active part in, these changes. A major issue for the protagonists in these processes—and an actual source for resistance—is how to express the problematic consequences of these developments so as to make these consequences exist as a shared problem, and be addressed (Labussière & Nadaï, 2018).

As suggested in the framing of this part of the book, any research work endorses a specific way of representing the issues that arise in relation with RET development, and it is then interesting to better understand the articulation between research practices and ways of conceptualizing social acceptance. This chapter ventures in this direction by looking back at our own research practice in analysing French onshore wind power development over the past fifteen years. Our research work is embedded in what we may call, referring to Bruno Latour, a politics of things (Latour, 2005a, 2005b). Much of our effort has been targeted at understanding the opposition to RET in relation to the politics that underlie their developments by following, describing and acknowledging the way in which many entities—humans but also non humans—are embarked and transformed in the process of developing these technologies. In what follows, we try to reflect on the articulation of this research practice with the conceptualization of social acceptance, to posit our successive problematizations of social acceptance within the academic field, and to a certain extent, to reflect on the representation of the dynamics of this academic field.

We first examine the factors that have influenced the course of our research work (§.1), we retrospectively distinguish between three ways in which we have addressed acceptance issues, which we successively present (§.2 to 4) before discussing this evolution in relation with the academic field of social acceptance of RET (§.5). This reflection points at a reformulation of the notion of social acceptance in order to address the web of interferences[1] at play in energy transitions processes. It also points out that pretending to do so challenges our position as social scientists in taking part into the field, the temporality and the reciprocity of fieldwork. Hence, challenging the notion of social acceptance is not enough. Shifting to interferences is just but a point of departure towards more engaged research practices.

Analytical Threads

Different factors may influence the way in which we, as researchers in social science, construct our research questions and programmes and, henceforth, frame our conception of people's relation to new energy infrastructures. In this section, we sum up the evolving French political context regarding the development of wind energy in France, under which we developed our research programme since the year of 2005; we point to a few factors, which we think influenced the course of our research programme; and propose to sketch this programme through three analytical threads.

Our research programme on French wind power started with the analysis of the first French parliamentary debate (Nadaï, 2007a) concerning this development and the rising opposition to it, five years after the adoption of national feed in tariffs for renewable energies. We then followed during ten years the evolution of this policy. We emphasized an inherited centralisation of energy policy and infrastructures, an early economic policy framing—feed in tariff being its pivotal instrument -, a reliance on private wind power developers, and a defiance of national politicians towards the capacity of local authorities to regulate the spatial development of renewable energies (Labussière & Nadaï, 2015). It was only at the end of the last decade—marked with a major concertation process about environmental issues, the 'Grenelle de l'Environnement' (2009)—that RET development became part of the national energy objectives. All of this contributed to a lack of articulation of renewable energy development with local territories and to an early local opposition to it. On average, about half of the administrative decisions about French wind power projects have been opposed to in court, either by developers (when administrative authorisations were refused) or by opponents (when administrative authorisations were granted). Acceptance issues have gained political recognition since 2005 in France.

Attempts at opening this development to political processes on a local scale through spatial planning (Wind Power Development Zones) had a short life (2007–2013). Early on, this device was pointed at by developers as an additional barrier to wind power development. It was also sued in court by opponents in an attempt to stop the development of wind power, before being eventually withdrawn from the national policy package under the pressure of green and RET developers' lobbies.

In 2015, the energy transition law (FR, 2015) marked a turn in French energy policy, by invoking the key role of local authorities and territories in the development of climate energy transition, and by opening possibilities for local actors to step on the financing of local energy projects. Financial participation and citizen investment became recognized by policy makers as potential means to foster the local acceptance of renewable energy projects. As a result, renewable energy acceptance is mostly addressed at the current time in terms of financing of renewable energy projects, with different options being considered: crowdfunding, citizen investment, and the development of RET projects by local authorities.

Issues of social acceptance of renewable energy projects have thus been framed in different ways by French policy makers—through economic incentives, local planning, regional planning, participative financing.

As researchers in social science who were partly funded by public policy programmes, we faced these evolutions both as issues and as political framing. While most of the calls for research under which we got financing left us pretty free of our analytical framing, the academic requirement to regularly—hence, swiftly—publish has probably been the most impacting requirement on our explorations, as it imposed a pace that did not necessarily allow time for deep critical reflexivity. Such reflexivity had to be gained and constructed through time, along opportunities for publishing, and with the hint of academic networks, and fieldwork encounters. Opportunities for publishing often had to be seized in the flow and came as moments of guided re-elaboration of already undertaken case studies, as crystallization of understandings. Academic and theoretical affinities have been decisive, if only because they steered encounters.

As far as our past publications about French wind power development witness it, our academic affinity has been with what we could name relational approaches—i.e. approaches that do not take individualities or entities as given and stable, but rather as in-the-making through processes and relations. This meant to follow, describe and acknowledge the way in which processes of developing renewable energies embarked entities—humans and non-humans—and changed ways of relating with / among them. While this resumes an enduring line of concern in our analyses of RET development, the theoretical strands we conveyed in the analysis have evolved. Some of them have been key references along the way, such as Akrich (1989), Latour (2005b) or Simondon (1989) relational approach to technology. Others have being called for in specific works or on specific objects such as non-representational approaches to analyse the spatial/

landscape planning of RET (Deleuze, 2005—map/tracing, sign; affect/ ethos) and pragmatist sociology to analyse the democratic dimension of energy transition processes (Dewey, 1946—public, problematicness; Marres, 2007, 2012—issue making, material participation).

In the end, this line of works converged with ***critical strands*** to both social acceptance (challenging the idea that opposition to RET should be reduced/ overcome—Batel, 2020)—and the energy transition (challenging the democratic dimension of energy transition processes—Labussière & Nadaï, 2018), as well as with a *co-productionist* agenda according to which the social and the sociotechnical are jointly produced (Chilvers & Kearnes, 2016). However, the many factors influencing our analyses and the particular organization and pacing of academic production, make it impossible to delineate clear chronological sequences in the work: publication do not necessarily follow fieldworks in time; occasions for publication (e.g. thematic issues) may influence the framing of case studies; fieldwork encounters can impact the framing of older case studies in new publications … Rather, we distinguish between three threads of analyses, respectively interested in: i/ the role of planning / permitting procedures and landscape policy in wind power development; ii/ the way in which different entities, humans or non-humans, are acknowledged (or not) in the development of renewable energy projects; iii/ the way in which resources, space and time are constructed in energy transition processes, resulting in certain political framing of—and a democratic deficit in—energy transition processes.

While this is based on reviewing our own work, it of course partly echoes results that have been pointed out in social acceptance research with different frames and terms (Ellis & Ferraro, 2016). The purpose in this chapter is to both point out the factors having influenced our research practices and the articulation between these practices and our evolving conception of social acceptance.

In what follows, we successively present each of these threads and the way in which they frame tensions around wind power development (§.2 to §.4), before discussing the ways in which social acceptance has been (implicitly) thematized and displaced in our work as well as the relation of our work to the academic field of sociological analysis of social acceptance.

Thread 1: Acceptance as a Consequence of Spatial Planning (Representations—Milieu)

This line of analysis is interested in the role that spatial planning plays in the emergence of tensions around wind power development. It has been initiated in the early phase of our research work, partly because of our background—one of us had a background as a geographer with an interest in the role of aesthetics and forms in spatial planning (Labussière, 2011), the other as a landscape architect with a recent interest in STS and the emergence of form in landscape projects (Nadaï, 2007b)—partly because the development of French wind power was triggering landscape, heritage and planning issues (Cf. §. 1).

Early field encounters were decisive in steering our attention to such issues. In 2007, the French Ministry of Culture had gathered in Paris, for a crisis meeting, all of its local branches for heritage protection. Wind turbine threats on heritage landscape, because of their far reaching visual presence in the countryside, had become a key issue. The administrative officer in charge of the protection of the Chartres Cathedral (Eure et Loir, France) presented a bright red urchin-like scheme figuring the dramatic State failure to protect the cathedral (a national heritage, at the centre of the urchin) from the visual interferences of wind turbines (located as dots in radiating red cones). The dotted red radiating scheme was materializing both the visual governmentality of French landscape (cones radiating for the cathedral, to be protected from visual intrusion) and its failure to govern (the decentralized presence of wind turbines in the yet unqualified landscape of the countryside). This setting was so telling that it triggered, for us, an in-depth, theoretical and empirical, inquiry into the making / unmaking of what we ended up naming a 'State landscape'—a formal, visual and centralized administrative tradition of landscapes protection, that was challenged by the emergence of wind power (Nadaï & Labussière, 2015). Why and how this was so, became a research concern. A parallel fieldwork, in the Aveyron department (Nadaï & Labussière, 2009), also witnessed the failure of normative planning processes. Administrative top down sieve map planning in South Aveyron had steered wind power developers towards areas judged less sensitive as to their heritage or environmental dimensions. This raised local concern about what was perceived as a spatial / environmental disqualification, if not injustice, and triggered the emergence of a local opposition to wind power in the concerned areas.

Working at the crossroad between critical geography, landscape studies and STS, we followed a 'constructive' approach (Latour, 2003) attentive to the way in which landscape/space was represented and circulated (graphic forms, material supports) in spatial planning. The focus on the relation between the material (un)making of landscape in planning and the rise of local opposition to wind power, was a way to point out factors behind opposition to wind power ('bad' planning). It also shed light on the relation of co-emergence between the technology and its human-non human 'milieu' (Akrich, 1989; Simondon, 1989). This consideration was progressively sharpened by analysing a case of successful wind power planning (Narbonnaise, South West France) (Nadaï & Labussière, 2013), which had allowed for a rebound in wind power development, after a local moratorium had been put on this development because of local tensions. The case study proved that, symmetrically, a potential for change and wind power development could emerge from planning processes that were attentive to the milieu (rather than to a predefined norm). This analytical line was further amplified by comparing case studies and emphasizing the importance of weaving together current socio spatial configurations with inherited ones (Labussière & Nadaï, 2014).

Along this thread, the underlying narrative is that social acceptance, even if not explicitly flagged in these analyses, resulted from reciprocal alignment between the technology and its milieu through attentive enough landscape planning processes or RET policy frameworks. While dialoguing within an academic strand interested in the role of spatial planning (Strachan et al., 2012), space and place (Devine-Wright & Howes, 2010) for wind power development—and thus pertaining to a "criticism" of social acceptance (Batel, 2020)—this thread of analysis, was also feeding a co-productionist agenda (hence 'critical'), be it only because of its reliance on Simondon notion of milieu.

Thread 2: Acceptance as Part of the Construction of a Relational Potential (Actants—Milieu)

The second line of analyses has been more directly interested in the fate of those which/who are embarked in the processes of wind power development. Rather than looking at the work of (mis) alignment through certain mediations—such as planning and landscape representations—the analysis aimed at targeting more directly those humans or non-humans that were affected and contributed to the (non) development of specific wind power projects: birds in one case study, villagers in the other.

Once again, this line of analysis has been strongly suggested by field encounters. In 2006, as we were analysing the innovative wind power planning process in the Narbonnaise, we interviewed a local bird watching NGO, which had just undertaken field experiments with a wind power developer. The project was located in a European migratory corridor at the French-Spain border on the Mediterranean shore (Languedoc-Roussillon, France). It involved a collaboration between the local branch of the French bird protection NGO (LPO) and a wind power developer (Nadaï & Labussière, 2010) in order to adapt the siting of an existing wind farm (repowering) to bird migration. Existing wind turbines were turned into a type of lab-scape: bird watchers hid behind the turbines, observed and followed individual birds in their crossing through the existing wind farm, gauging their individual cognitive/strategic ability in flying through it. Both the reshuffling of bird classifications (according to crossing ability rather than to statutory protection) and the drawing of individual bird trajectories allowed for the mapping of migratory micro-corridors, which paved the way for wind farm siting proposals judged compatible with bird migration. By focusing on individual stories so as to capture birds' intelligence, this experiment endowed birds with new capacities and opened a new potential for sharing the wind. In this process, entities were requalified and performed differently: birds became more skilled (wind farm compatible) and the wind farm became more compatible with migrating birds. By focusing on human—non human relations (affect, animal charisma), the analysis brought forth the ontological dimension of the process and its potential to give rise to renewed compatibility between wind power development and bird protection (human-non human assemblage).

The second case study followed the progressive structuring of a network of opponents to a local wind power project in a village located in the Parisian basin (2003–2009) (Nadaï & Labussière, 2017). The analysis traced issues as they emerged in relation with the project and its interferences with shared landscape resources (a local 'plain') that were collectively managed (land consolidation association) and used at the village level (Sunday walk). Relying on pragmatic sociology of issue making and public formation (Dewey, Marres), it followed a collective of opponents in the making. It showed the ways through which part of these villagers, which at the outset were not opponents to wind power per se or to the project, progressively became opposed to the way in which it was developed. We showed that the incapacity of permitting and policy institutions to acknowledge the collective trouble raised by the project (unravelling of village life, divisions within families, stalling of social activities ...) mirrored a narrow framing on renewable energy resources, as wind speed and private land. This case study thus pointed at a failed wind power potential: opposition to wind power resulted from a lack of attention to relations, and especially to the collective dimension of certain resources.

In both case studies, the focus was on the construction (or lack thereof) of specific relations. Both case studies foreground an ontological dimension of wind power development processes: identities, roles and capacities to influence the course of RET project development are distributed by and through these processes. The case studies flag relational potentials for change in the (un) making. Acceptance (issue) is thus conceived as the outcome of the construction (or lack thereof) of this relational potential. As much as thread 1, this thread is both 'critic' and 'critical' to social acceptance. Yet, it is much more explicit in its ontological lessons, because it makes clear the transformations that entities undergo through RET development processes and the related potentials for contribution (or not) to these processes they are given (or not). In so doing, such a perspective echoes analyses in the field of participation, which emphasize the construction of publics by participatory settings (Laurent, 2011; Lezaun & Soneryd, 2007).

Thread 3: Acceptance (Issue) as an Expression ('Problematicness') of Interferences

The third thread of analysis, more recent, has built on the former two. It was very much triggered by fieldwork encounters in Aveyron—meeting with overwhelmed opponents—which pointed out a type of democratic deficit in the way RET was developed. Democratic deficit meant here that the parties affected by this development did not have adequate access to the arenas in which the decisions about this development were made. Moreover, both the Aveyron and the Seine-et-Marne case studies pointed at the relation between the way in which the energy resource was framed as naturally given (wind speed) and appropriable by anyone who could invest in / develop a technological artefact (turbine, solar panel...), regardless of the collective dimension of other resources engaged (land or landscape uses, effort spent in coordinating shared visions about where to develop these new energies …), and the rise of acceptance issues.

Building on pragmatic thinking, in particular that of John Dewey (1946) about 'publics' as political moments and collectives, and its extension to issue making and material participation (Marres, 2007, 2012), the idea was to focus on the relationships between *all* the entities convened (intentionally or unintentionally) in energy transition processes and explore the transformative (or disruptive) scope of the corresponding relational assemblages. Building on a large set of case studies, we proposed to distinguish between transition *trajectories* and the *processes* that underlie them and condition their deployment (Labussière & Nadaï, 2018). "Trajectories" are devised in advance, through actors and categories of action that have been formed and stabilized—often involving modelling exercises and technology based visions in the field of the energy transition. Different from this, "processes" refer to a multiplicity of entities that participate in the construction of these trajectories, sometimes at the level of RET projects, without necessarily having control over them.

For instance, the development of a new wind farm may cause discontinuities in experiences (e.g. impeded animal migrations, fragmented agricultural communities, distant co-visibilities and changed perception of heritage elements) which, if not perceived and dealt with, give rise to potential inequalities. Building on thread 2, the analysis thus approaches the transition as a period of "ontological trouble" according to which not all the entities involved in the processes are called upon to "transition" with the same opportunities for being taken into account (Cf. the contrast between the Narbonnaise and Seine et Marne case studies). These entities find different existences in energy transition processes according to the degrees to which they are concerned, impacted, involved, acknowledged and redefined with or without their consent. Following Noortje Marres, such an approach to potential endows the collective qualification of problems ('problematicness') with a central role in the formation of material 'publics' and of their capacity for action.

Based on this observation, the challenge is no longer to operationalize trajectories in order to reach predefined potentials. It is to acknowledge and assess the "interferences" that these trajectories generate, which in turn allows for the analysis of the (non) emergence of shared transition potentials. The question of "interferences" thus becomes the focal point of the inquiry and the concern for symmetry—that is, acknowledging all the parties convened into the process, addressing them with the same analytical lens and giving the same status to failed and successful options—is the analytical key to understanding the democratic dimension of the process.

Under such an approach, acceptance (issue) endorses a broader and deeper meaning as it is the expression ('problematicness') of interferences or, said differently, the expression of a democratic potential (deficit) in the conduct of the energy transition (Bell et al., 2013; Labussière & Nadaï, 2018). Most importantly, however, while acceptance still relates to alignment and ontological work, turning to interferences broadens the scope of the analysis. Indeed, analysing interferences challenges as well the construction of resources, space and time in energy transition processes, because this construction underlies the potential for participation and change.

This approach also has two major implications for thinking about the energy transition. Firstly, it recalls us that energy is always more than just a matter of energy; it questions just as much our relationships with biodiversity, landscapes, or spatial planning, and through them the handling of new tensions between development, justice and the environment. Secondly, relationships—which are at the heart of pragmatic inquiry—constitute a critical resource for the transition in that their collective specification and qualification makes it possible to support solutions that bring about social and environmental justice. Such conclusions echo the co-productionist agenda and its recent call for a just energy transition (Jasanoff, 2018).

Discussion: Interference as Hint for Different Encounters

This chapter proposed to undertake a reflective exercise about our preconception of social acceptance by considering our research practice about the development of wind power and energy transition processes over the past fifteen years. Attempting to trace back the ways in which we framed the tensions around RET development in France, we have shown that several factors influenced it—including our backgrounds, our academic networks and, importantly, fieldwork encounters. As one of the authors has written about, the field has a potential to resist our research framing and programming, and the capacity to entice us to explore new tracks (Labussière & Aldhuy, 2012). As far as we are concerned, several encounters have been decisive in the way we evolved our questioning about wind power development.

Important also has been the process of production of research papers, often resulting from academic network opportunities, which enticed us to re-interpret case studies along specific concerns, and makes it difficult to ex-post delineate a clear temporal succession in our research work. We thus distinguished between 'threads of works' mirroring the way in which we felt our reflection about the tensions raised by French RET development had progressed.

Each of these threads foregrounds a different articulation to which to relate social acceptance and its emergence (or lack thereof): representation and milieu, actants and milieu, interferences. Hence, each of these articulations points to a different problematization of social acceptance: acceptance (issue) as a result of reciprocal alignment between the technology and its milieu through attentive spatial planning processes; acceptance (issue) as the outcome of the construction (or lack thereof) of a relational potential; and acceptance (issue) as the expression ('problematicness') of interferences.

As we have shown, each thread is embedded in a definite analytical strand, at the same time that it is both 'a criticism' and 'critical' to the notion of social acceptance. In the end, even if the 'critical' perspective has been said to challenge the relevance of RET development (because of its consequences) when the 'criticism' pointed out factors of resistance to it (Batel, 2020), it seems difficult to clearly sort out these perspectives here. The reason for this is that factors of opposition ('criticism' perspective) most often result from the ontological consequences of RET development, so that, at least in our research work, pointing out these factors challenges RET development by shedding light on its social and political consequences (discriminations, injustice and lack of democracy). Again, this may be singular to our research practice, and relate to our embedding in an academic field (STS, critical geography) that challenges ontologies (hence, the premise of individuality). As emphasized by Batel (2020), the 'criticism' approach to social acceptance, has otherwise significantly been developed by (social, environmental) psychology, a discipline that works within this premise.

Each thread in our research work echoes works and results pertaining to the academic field of social acceptance. This especially is the case if we include in the definition of this academic field, the works concerned with public participation and the coproduction agenda (Chilvers, Laurent,

Jasanoff, Marres). The co-production agenda, though, has several scales of interpretation and operationalization: it points to both the co-production of orders (social and technological) and objects (technologies, individualities). As far as the above witnesses, our research work has mostly proceeded from the latter, through detailed case studies, in order to point to the former; we did build from Laurent and Marres framing, but not from Chilvers or Jasanoff ones.

Reformulating acceptance issues as the expression of interferences in energy transition processes, however, should not be regarded as a closure within this agenda. As far as we are concerned with the urgency of environmental issues, following interferences calls for enlarging the requirement for symmetry so as to include ourselves (the observer). Interferences are not only among and for the others. They cannot be formulated in terms of distant options or alternatives. Following their making calls for attending to the multiple arenas in which ontologies and their politics are constructed and at play, *including that of the social scientists themselves*—a point that Noortje Marres has made very clear with the notion of experimental ontology (2013). Thus, emphasizing the pragmatic heritage of the co-production agenda carries with it a *methodological displacement that goes beyond the critical perspective and a new departure point in how we address these issues* (Cf. Table 8.1). The point is no longer to challenge the relevance of RET development by only shedding light on its underlying politics and consequences ('critical perspective'). It also is—by including ourselves as sources and ends of interferences—*to reflexively engage with the field and be challenged by it as social scientists, because this paves the way to enlarged experimentations and ontologies, and thus to new possibilities.*

Pretending to follow interferences, challenges our position as social scientists, the method and the material devices we recourse to in taking part into the field, as well as the temporality and the reciprocity of fieldwork. While such a change has been partly thematized under the headings of material participation and experimental ontology (Marres, 2013; Labussière & Nadaï, 2018, pp. 22–30 for an application to energy transition issues) much remains to be done about how to operationalize such a change. Interestingly, and echoing to the role of fieldwork in enticing methodological changes, recent research works about what has been called 'research fatigue' in the field of energy transition research come as a warning (Walsh et al., forthcoming). 'Research fatigue' points out situations in which actors in the field refuse to collaborate with social scientists because they no longer believe in the ongoing politics of 'the' energy transition

Table 8.1 Overall displacement from 'pre-conception' to 'academic engagement'

Research implication	*Interferences*	*Approaches to acceptance*
From pre-conception	The 'technology' and the 'social' pre-exist to their "interactions"; "interferences" are not made visible	Technology as a potential for change, social as a potential for resistance; "acceptance" means taming social passions; the researcher is an external observer
Thread 1: representations of landscape and practices of wind power planning	Lived landscapes and local uses ill-named and ill-represented by practices of mapping; some cases of innovative planning opened to local concerns	"Acceptance" is the result of reciprocal alignment between the technology and its milieu
Thread 2: ontological issues of wind power processes	Finding out new relational potentials (e.g. changing bird observations / knowledge, micro-siting of turbines) or not (e.g. ignoring landscape commons)	"Acceptance" is the outcome of the construction / taking into account (or not) of relational potentials
Thread 3: multiplicity of entities convened in energy transition processes	Capacity to acknowledge the various entities concerned, impacted, involved and redefined with or without their consent in a process of transition	"Acceptance" as the expression ('problematicness') of interferences
To academic engagement	Following the multiple arenas in which ontologies and their politics are constructed, including that of social scientists	Making interferences explicit is not a closure nor a remedy but opens a potential for collaborative experiments

(our experience in Aveyron), or because they do not see the value of the usual talk-leave-and-publish research sequence of social scientists, and probably because of both. Of course, 'research fatigue' is not just a symptom manifested by an external body-field but makes ourselves part of the problem. It opens a more general questioning about how we articulate both the field and a global knowledge economy. Hence, what 'research fatigue' also carries with it, is a call for a different relation and engagement of social science researchers with the field. Collectively experimenting ways to acknowledge interferences and support emerging publics (e.g.

collaborative research, art/sciences experiments) may usefully challenge and invigorate social sciences to contribute to the setting up of more just processes of energy transition—a challenge that is ahead of us, and to which some works already call us to engage with (Marres et al., 2018; Segers, 2019; Wilkie & Michael, 2018).

Note

1. The notion of "interference" brings the emphasis on emerging relational realms induced by energy transition processes. We distinguish between four notions: "interaction" and "interference", "relationship" and "interrelationship". To put it shortly, the notion of interaction pertains to analytical strands—such as system analysis—which presuppose the existence of distinct entities and analyse the set of relations between these given entities. Different from this, the notion of interference points out situations in which the parts do not necessarily precede their being part of a common reality. In such situations, the relations are constitutive of the parts, which ontologies then depend on the set of relations at work. Importantly, the becoming of interferences is highly uncertain, as they can evolve towards reciprocal modes of relationships acknowledging the consequences of the process for all the parties—which we term "interrelationship"—or towards ill-related and sometimes conflictual reality.

References

Akrich, M. (1989). La construction d'un système socio-technique. Esquisse pour une anthropologie des techniques. *Anthropologie et Societes, 13*(2), 31–54. https://doi.org/10.7202/015076ar

Batel, S. (2020). Research on the social acceptance of renewable energy technologies: Past, present and future. *Energy Research & Social Science, 68*, 1015–1044. https://doi.org/10.1080/1523908X.2017.1417120

Batel, S., Devine-Wright, P., & Tangeland, T. (2013). Social acceptance of low carbon energy and associated infrastructures: A critical discussion. *Energy Policy, 58*, 1–5. https://doi.org/10.1080/1523908X.2017.1417120

Bell, D., Gray, T., Haggett, C., & Swaffield, J. (2013). Re-visiting the 'social gap': Public opinion and relations of power in the local politics of wind energy. *Environmental Politics, 22*, 115–135. https://doi.org/10.1080/09644016.2013.755793

Chilvers, J., & Kearnes, M. (Eds.). (2016). *Remaking participation: Science, environment and emerging publics*. Routledge.

Deleuze, G. (2005 [1968]). *Différence et répétition.* Presses Universitaires de France.

Devine-Wright, P., & Howes, Y. (2010). Disruption to place attachment and the protection of restorative environments: A wind energy case study. *Journal of Environmental Psychology, 30*(3), 271–280. https://doi.org/10.1016/j.jenvp.2010.01.008

Dewey, J. (1946 [1927]). *The public and its problems. An essay in political inquiry.* Gateway.

Ellis, G., & Ferraro, G. (2016). *The social acceptance of wind energy: Where we stand and the path ahead.* Publications Office of the European Union, Luxembourg. https://doi.org/10.2789/696070

FR. (2015). Loi n° 2015-992 du 17 août 2015 relative à la transition énergétique pour la croissance verte (1). *Journal Officiel de la République Française, 0189*, 14263.

Jasanoff, S. (2018). Just transitions: A humble approach to global energy futures. *Energy Research & Social Science, 35*, 11–14. https://doi.org/10.1016/j.erss.2017.11.025

Labussière, O. (2011). Eléments pour une prospective du milieu. *L'enjeu du sensible en aménagement. Cahiers de Géographie du Québec, 54*, 499–515. https://doi.org/10.7202/1005608ar

Labussière, O., & Aldhuy, J. (2012). Le terrain ? C'est ce qui résiste. Réflexion sur la portée cognitive de l'expérience sensible en géographie. *Annales de géographie, 687–688*(5), 583–599. https://doi.org/10.3917/ag.687.0583

Labussière, O., & Nadaï, A. (2014). Unexpected wind power 'potentials': The art of planning with inherited sociogeographical configurations (France). *Scottish Geographical Journal, 130*(3), 152–167. https://doi.org/10.1080/14702541.2014.922210

Labussière, O., & Nadaï, A. (2015). Wind power landscapes in France: Landscape and energy decentralisation. In F. Marina, P. María-José, & N. Alain (Eds.), *Renewable energies and European landscapes. Lessons from Southern European cases* (pp. 81–93). Springer Netherlands. https://doi.org/10.1007/978-94-017-9843-3

Labussière, O., & Nadaï, A. (Eds.). (2018). *Energy transitions—A socio-technical inquiry* (Energy, climate and the environment series). Palgrave Macmillan. https://doi.org/10.1007/978-3-319-77025-3

Latour, B. (2003). The promises of constructivism. In D. Idhe (Ed.), *Chasing technoscience: Matrix of materiality.* Indiana University Press.

Latour, B. (2005a). From Realpolitik to Dingpolitik. In B. Latour & P. Weibel (Eds.), *Making things public.* MIT Press.

Latour, B. (2005b). *Reassembling the social—An introduction to Actor-Network-Theory.* Oxford University Press.

Laurent, B. (2011). Technologies of democracy. Experiments and demonstrations. *Science and Engineering Ethics, 17*(3). https://doi.org/10.1007/s11948-011-9303-1

Lezaun, J., & Soneryd, L. (2007). Consulting citizens: Technologies of elicitation and the mobility of publics. *Public Understanding of Science, 16*, 279–297. https://doi.org/10.1177/0963662507079371

Marres, N. (2007). The issues deserve more credit: Pragmatist contributions to the study of public involvement in controversy. *Social Studies of Science, 37*, 759–780. https://doi.org/10.1177/0306312706077367

Marres, N. (2012). *Material participation. Technology, the environment and everyday publics.* Palgrave Macmillan. https://doi.org/10.1057/9781137029669

Marres, N. (2013). Why political ontology must be experimentalized: On the eco-showhome as a participation device. *Social Studies of Science, 43*, 417–443. https://doi.org/10.1177/0306312712475255

Marres, N., Guggenheim, M., & Wilkie, A. (Eds.). (2018). *Inventing the social.* Mattering Press.

Nadaï, A. (2007a). Planning, Siting and the local acceptance of wind power: Some lessons from the French case. *Energy Policy, 35*, 2715–2726. https://doi.org/10.1016/j.enpol.2006.12.003

Nadaï, A. (2007b). Site ou l'émergence d'un paysage. *Cosmopolitiques, 15*, 121–134.

Nadaï, A., & Labussière, O. (2009). Wind power planning in France (Aveyron): From State regulation to local experimentation. *Land Use Policy, 26*, 744–754. https://doi.org/10.1016/j.landusepol.2008.10.018

Nadaï, A., & Labussière, O. (2010). Birds, turbines and the making of wind power landscape in South France (Aude). *Landscape Research, 35*(2), 209–233. https://doi.org/10.1080/01426390903557964

Nadaï, A., & Labussière, O. (2013). Playing with the line, channelling multiplicity: Wind power planning in the Narbonnaise (France, Aude). *Environment and Planning D: Society and Space, 31*, 116–139. https://doi.org/10.1068/d22610

Nadaï, A., & Labussière, O. (2015). Wind power and the emergence of the Beauce landscape (Eure-et-Loir, France). *Landscape Research, 40*, 76–98. https://doi.org/10.1080/01426397.2013.784732

Nadaï, A., & Labussière, O. (2017). Landscape commons, following wind power fault lines: The case of Seine-et-Marne (France). *Energy Policy, 109*, 807–816. https://doi.org/10.1016/j.enpol.2017.06.049

Segers, J. (2019). Tactics for tough issues. *Multitudes, 77*(4), 94–100. https://doi.org/10.3917/mult.077.0094

Shove, E. (2010). Social theory and climate change. *Theory, Culture and Society, 27*, 277–288. https://doi.org/10.1177/0263276410361498

Simondon, G. (1989 [1958]). *Du mode d'existence des objets techniques.* Aubier.

Strachan, P., Szarka, J., Cowell, R., Ellis, G., & Warren, C. (2012). *Learning from wind power: Governance, societal and policy perspectives on sustainable energy*. Palgrave Macmillan.

Walsh, K. B., Haggerty, J. H., Jacquet, J. B., Theodori, G., & Kroepsch, A. (forthcoming). Uneven impacts and uncoordinated studies: A systematic review of research on unconventional oil and gas development in the United States. *Energy Research and Social Science*. https://doi.org/10.1016/j.erss.2020.101465

Wilkie, A., & Michael, M. (2018). Designing and doing: Enacting energy-and-community. In N. Marres, M. Guggenheim, & A. Wilkie (Eds.), *Inventing the social* (pp. 125–147). Mattering Press.

CHAPTER 9

How to Assess What Society Wants? The Need for a Renewed Social Conflict Research Agenda

Eefje Cuppen and Udo Pesch

Introduction

The field of social acceptance of RET is basically concerned with the question of what society wants (or does not want) in relation to RET, or in other words, societal assessment. One of the key challenges for this field is therefore how to gauge societal assessment. This is indeed a challenge, because the notion of societal assessment is not unequivocal. It brings about questions such as, firstly, who constitutes the society, or 'the public'? Whose preferences or assessment needs to be included? There are no clear demarcations to decide who belongs to 'a public' and who does not. A

E. Cuppen (✉)
Leiden University, Leiden, The Netherlands
e-mail: e.h.w.j.cuppen@fgga.leidenuniv.nl

U. Pesch
Delft University of Technology, Delft, The Netherlands
e-mail: upesch@tudelft.nl

S. Batel, D. Rudolph (eds.), *A critical approach to the social acceptance of renewable energy infrastructures*,
https://doi.org/10.1007/978-3-030-73699-6_9

second question is how societal assessment is related to, or can be derived from the assessments of individual members. There are many ways to come from the assessments of individual members to a singular expression. For instance, the preferences of individual members can be aggregated, averaged or the majority rule can be applied to find out what society wants. However, it can also be said that society is more than the 'sum of its parts' and that a societal assessment should be derived from the involvement of all members. A third question is which claims are considered to be legitimate contributions to societal assessments. For instance, assessments that are based on self-interest or claims that contradict science-based expertise are often considered to be invalid.

Next to the more traditional ways such as elections and consultations, there is a wide range of possibilities that actors may deploy to form and forward their assessment. Particularly in the energy domain, such assessments may emerge bottom-up as self-organised forms of engagement (Cuppen, 2018), such as community initiatives that are involved in the production and distribution of energy, so-called prosumers. Societal assessment also takes place via consumer choices, as testified by the manner in which the deployment of PV at the household level has taken off in recent years (Walker & Cass, 2007). Moreover, citizens can organise societal assessment by opposing policies, plans and projects that affect them, that is, social conflict. Such social conflict on RET has become a widespread and recurring phenomenon worldwide.

In social acceptance literature, conflict is typically regarded as lack of acceptance, or, the opposite of social acceptance, producing risks that need to be mitigated (Aitken, 2010). This view implies that social conflict is something that needs to be "reduced", in order to "increase social acceptance" (e.g. Gross, 2007, p. 2727). Also among policy-makers conflict is typically seen as something to be avoided, as it may create obstructions to effective policy making. We argue that such a conflict-averse view is biased towards an instrumental, managerial view of what social conflict is and how it is to be engaged with. It aligns with neo-liberal views on community engagement (Batel, 2020; Renn & Schweizer, 2009), where conflict needs to be prevented or 'managed', and if it is engaged with, it is through a negotiation of interests, trading community interests with business or governmental interests. A good outcome then is one where there are 'mutual gains' (Susskind & Field, 1996).

Instead of social conflict being a risk that needs to be managed, we argue that social conflict is an important manifestation of societal assessment that is at par with conventional mechanisms of societal assessment such as elections, referenda or invited participation. In this chapter, we expand the notion of social conflict as a form of participation as posited by Cuppen (2018), by giving a theoretical refinement and by introducing a number of empirical illustrations from research in which the authors have been involved. With this, we aim to show what the notion of social conflict as societal assessment means for the question that is central in social acceptance literature as well as in RET policy and practice, which is, how societal assessment of RET could be gauged or approximated.

In the remainder of this chapter, we will first further conceptualise the notion of societal assessment. In the section thereafter, we identify and discuss three false assumptions about social conflict in social acceptance literature. We then move on to discuss the implications for research on social acceptance in the final section.

Social Conflict as Societal Assessment

Societal assessment is fundamentally intangible. It will and can never be determined what it actually is or even what it is about. Instead, it is a 'social imaginary' (Pesch, 2019; Taylor, 2002), building on a mental representation of what 'society' is or should be. Such an imaginary motivates societal actors to engage in collective deliberation about affairs that pertain to the public at large. This account of societal assessment comes close to the idea of 'rhizomes' introduced by Deleuze and Guattari (1987). Rhizomatic phenomena are best seen as processes which have no clear boundaries in time or space; as such, they cannot be separated from an external context. To some extent, this rhizomatic approach to social conflict can also be recognised in complex systems approaches to citizen involvement (Groves, 2010). In such systemic approaches also, the interdependencies between different practices, events and activities are emphasised (Chilvers et al., 2018; Mansbridge et al., 2012). Seeing societal assessment as a rhizomatic process implies that there are no clear beginnings, ends, directions or elements. This makes it hard to find points of departure that allow for analysis, understanding or management of societal assessment.

The vantage point of societal assessment allows for a different and much more constructive perspective on social conflict. It also does justice to the fact that actors *learn* –actors change or adjust their assessment through interaction with others and because of increased knowledge about the technical, institutional and political complexities of a certain issue. As such, social conflict can be said to have the potential to enrich the process of societal assessment, as has also been found in detailed case studies on societal energy disputes (Anderson, 2013; Usher, 2013).

Three Shortcomings in Social Acceptance Literature

The predominant instrumental account of social acceptance reproduces three incorrect assumptions; as such, its dominance needs to be challenged. First, the instrumental account denies that conflicts are a source of information about how society assesses a certain policy or project. With this, it fails to acknowledge that conflictual engagement may be used "to the broader advantage of low carbon societies" (Barry & Ellis, 2011) and that social conflict is an inherent part of democracy and political participation (Batel, 2018; Verloo, 2018). Social conflict can be considered a form of self-organised public engagement with regards to RET (Cuppen, 2018; Reed et al., 2018). We will elaborate on this in the section 'Social Conflict as Participation'.

Second, this account assumes a binary opposition between members of the general public (usually seen as lays) on the one hand, and professionals and experts on the other hand (Aitken, 2009; Pesch et al., 2020), or between proponents and opponents as two artificial hegemonic blocs (Barry & Ellis, 2011). Social conflict however is better portrayed as a dynamic process of socio-political interactions among a wide range of actors: citizens, project developers, governments, civil society groups, supervisory bodies, and so on. It is not only responses by citizens that matter for understanding how social conflicts evolve, but also responses and anticipations by for example, CEOs or communication officers of energy companies, civil servants of municipalities and so on(cf. Walker et al., 2011). We will elaborate on this in the section 'Social Conflict as a Multi-actor Process'.

Third, the predominant account of social acceptance associates matters of social acceptance with discrete projects or policies, in the sense that these projects and policies can be studied and managed as independent activities not influenced by events taking place elsewhere. However, social conflicts on RET interact with social conflicts that have occurred in another time or place or on another technology (Cuppen et al., 2020) as well as with other forms of public engagement. Attempts of invited participation may for instance trigger community protests, or protest against (lack of) policy plans may motivate citizens to engage in energy communities (Pesch et al., 2019). We will elaborate on this in the section 'Social Conflicts Interact with Other Conflicts'.

Social Conflict as Participation

In policy and planning, participation of citizens is usually proposed as a means to prevent or deal with conflict, or to legitimise decisions already taken (Bogner, 2012). In this, participation is typically contrasted to a top-down or expert-analytic approach (Stirling, 2008; Wynne, 2001) that is not inclusive enough and not considerate of specific wants and needs of citizens and thereby easily provokes conflict. In that way, participation is conflated with invited (Wynne, 2007), or orchestrated (Leach & Scoones, 2007) engagement.

Although there are different rationales (Fiorino, 1990) and tools (Hisschemöller & Cuppen, 2015) for participation, most of the approaches are implicitly or explicitly building on Habermas' idea of an ideal speech situation. In such an ideal speech situation, citizens can express their viewpoints without interference of power asymmetries (Renn et al., 1997). It is a safe and egalitarian situation where "actors seek to reach common understanding and to coordinate actions by reasoned argument, consensus and cooperation rather than strategic action strictly in pursuit of their own goals" (Habermas, 1970, p. 86). Most participatory approaches take this notion to assume that consensus is required to achieve progress in decision-making processes (Leeuwis, 2000).

However, as mentioned, social conflict on RET can also be seen as a form of participation (Cuppen, 2018). In processes of social conflict, actors form their opinions and views about the RET, as well as about the decision-making process, the institutional structures, and so on. As such, social conflict is a form of societal assessment, that can be used productively as a possibility for actors to learn.

We observed such learning through social conflict in the case of the Dutch interest group for residents living near (planned) wind farms (NLVOW).[1] This interest group was founded in 2013 to give a voice to the interests of residents during decision-making and construction of wind farms and turbines (www.nlvow.nl). Although the initiators state that the association was not founded because its members were against any specific form of energy production, they were in the beginning mainly active to support protest. Over time, the association changed its profile, for example, since 2018 they have also focused on other 'new' forms of energy production such as biogas, geothermal and solar energy. They also adopted a more constructive approach towards government, for example, jointly taking part in projects or advising local and national governments and project developers on how to involve citizens. Over time, the association realised that the main interest that they want to defend, namely the right of residents to have a say in governmental policy and planning, goes beyond RET. This is when two members started a new organisation: the Dutch Platform Citizen Participation and Governmental Policy (NPBO; www.npbo.nl). This example shows, firstly, that conflict is indeed a process of societal assessment in which actors form their opinion and mobilise voice to express their values and preferences. Secondly, it shows that learning occurs amongst the actors involved in social conflict. The initiation of a new organisation for citizen participation is an articulation of the fact that these actors realised that they do not only care about visual impact or noise of wind turbines, but also—and perhaps even more prominently-about democratic values such as justice and legitimacy.

A second example is the social conflict on natural gas production in the Netherlands. The Groningen gas field is one of the largest reservoirs in the world. Since production started in 1963, it has become very important for the Dutch economy and domestic energy supply (van der Voort & Vanclay, 2015). The region experienced minor earthquakes over the last decades, but only after an earthquake of 3.6 on the Richter scale, the earthquakes gave rise to debate amongst residents and politicians (van der Voort & Vanclay, 2015). The tremors resulted in various impacts, such as damaged properties, declining house prices and safety hazards. The societal assessment expressed through this conflict, however, is broader than that. Residents expressed concerns about the institutional structures of gas production, in particular the role of the Dutch Petroleum Company (NAM; a partnership between Shell and Exxon Mobile) and the Dutch government. It became a socio-political conflict, in which justice featured explicitly as an important theme. Approaching the conflict as societal assessment thus highlights the wide range of themes, issues and values that residents, and later also other actors, assess relevant and important for decision-making on natural gas.

These examples show that seeing societal conflict as a mere risk ignores the possibility that conflicts forward a societal assessment, in the sense that a wider range of values, issues and concerns are made explicit, which may subsequently be used to improve the democratic and substantive quality of decision-making. The reluctance to attend to such values and concerns may contribute to the further counterproductive degradation of societal support, fuelling societal resistance. Furthermore, it denies the possibility that actors can change their assessment as they become better informed about the nature of the project, policy or institutional structures.

Social Conflict as a Multi-actor Process

Much of the social acceptance literature focuses on publics (Wolsink, 2019); that is, those groups in society that 'need to accept' a certain phenomenon. Those publics are usually considered to have a preference that can be determined as a singular measure, and composed out of people that only have non-professional interests (Pesch, 2019). However, looking at social conflicts on RET, it becomes clear that societal assessment is a dynamic *process* in which groups and their preferences are not fixed.

In this dynamic process, expectations play a formative role as they shape interactions between publics and project developers (Walker et al., 2011). These expectations not only have to do with the RET, but also with the roles, responsibilities and capacities both of themselves and of the actors they are engaging with. In Science & Technology Studies such expectations have been studied under the label of 'imagined publics' (Barnett et al., 2012; Walker et al., 2010), to explain how publics are anticipated, perceived, and reacted to by science and technology actors (Marris, 2015; Welsh & Wynne, 2013; Wynne, 2006).

It is precisely because of the performativity of imagined publics and expectations, that the conventional image of public acceptance should be scrutinised. After all, imagined publics can give rise to self-reinforcing patterns of interaction, for instance by perpetuating beliefs about NIMBYism (Bell et al., 2005; Wolsink, 2000).

Because interactions between citizens and government or firms who are planning RETs are so key to societal assessment, it is striking that there is so little attention to the behaviour, expectations and imaginations of governments and firms when it comes to the study of social acceptance. There are a few valuable contributions that give insight into how governmental and industrial actors view publics and public engagement, such as Cotton and Devine-Wright (2012), Van de Grift et al. (2020), Devine-Wright (2011), and Burningham et al. (2015). If we really want to understand processes of social acceptance, we need to understand, firstly, how *all* actors engaged in these processes behave and strategise, not just citizens or residents (Walsh et al., 2017). Secondly, we need to understand how anticipations and strategies of all these actors blend together in interactional spaces, producing specific responses and decision outcomes. This is in line with Wolsink's (2019) call for more attention to institutional dimensions of social acceptance, as well as avoiding a bias towards *public* acceptance, and instead also look into *socio-political* and *market* acceptance (Wüstenhagen et al., 2007).

A starting point for eliciting the expectations, assumptions and imaginaries that actors deploy is to find out how actors within organisations develop a conception of their own role in the interaction with others in RET planning. For instance, we have found that so-called community engagement professionals, the professionals responsible for engaging communities in the development of RET, typically maintain three perspectives upon their own role. A first group sees themselves as intermediaries between their organisation and the community; a second group sees community engagement as an inherent part of project management, aiming to remain in control of the process; and a third group sees community engagement as something they do because they want to comply with laws and regulations—equating legitimacy with legality (van de Grift et al., 2020).

Likewise, it is highly important to understand how interactions take place between organisations. These interactions may point us towards tensions and ambiguities given the tasks and responsibilities these actors have to cope with. We observed the importance of such interactions in a case of a wind turbine park in the Dutch province of Groningen. The national government had given the provincial authority a directive on the to-be-installed capacity, without instructing how this should be implemented. To accommodate this directive, projects had to be developed, and the provincial authority allocated locations for these projects. Menterwolde was one of the municipalities whose local authorities were confronted with this top-down planning. At the same time, this municipality had to deal with the escalating resistance from its local communities against wind parks, and specifically from the village of Meeden. Residents did not merely contest the wind turbines, but also unleashed, sometimes violent, campaigns against farmers who leased their land to the corporate developers. In other words, the municipality had to mediate the demands of the local public who oppose wind parks, it had to protect the threatened farmers, while also following the directives of the provincial and national authorities and adhering to their legal responsibilities as municipality. The municipality was caught between fires, not able to establish or follow a unitary preference or interest.

Social Conflicts Interact with Other Conflicts

Social conflict and social acceptance of RET are typically studied by drawing the case boundaries around a project or technology. However, many, if not most, conflicts, are shaped by other conflicts that have occurred in another place, at another time, or for another technology or policy issue. In other words, a focus on cases that are separated to serve analytical purposes ignores the way in which social conflicts are informed by other social conflicts—we have termed these dynamics as 'controversy spillover' (Cuppen et al., 2020). For instance, shale gas protests in the Netherlands were inspired by the ecological problems that ensued shale oil exploitation in the United States and earthquakes that happened in the United Kingdom due to shale gas drilling, and a recurring claim of protestors in the wind energy controversy in Groningen was that the province is again treated as an 'energy colony' by other parts of the country, as for centuries the region has been exploited for peat, natural gas, and now wind. Spillovers may both reproduce and qualify the notion of 'place attachment', that is, the special meaning that residents reserve for the social and natural environment in which they live (Devine-Wright, 2013).

The idea that social conflicts on energy infrastructure interact with social conflicts is not new. However, it is typically studied as 'context'. We have argued that these spillovers deserve to be analysed as objects of analysis themselves (Cuppen et al., 2020). After all, little is known about how issues travel and what their impact is on the process of societal assessment. To answer this question, we may link back to the way interactions are shaped by expectations. After all, it seems reasonable to suggest that actors will substantiate these expectations by looking at other events and experiences. Actors will have to get an idea about how a project and policy may work out. They can do so by looking at other conflicts to learn things like what can be considered relevant impacts, risks and concerns; what kind of promises have been made or broken? To find an answer to such questions, they may search the internet, or contact activist groups elsewhere to get more knowledgeable. As such they can acquire information about other RET conflicts and experiences and expectations may transfer from one case to the other. It needs to be emphasised that not only members of the general public recruit information from events that have taken place elsewhere to make sense of a certain issue. Also, project-developers, industry and public authorities shape their strategies based on the experiences and events that took place elsewhere.

Implications for Research on Social Acceptance

This chapter has sketched out how social conflicts have wider ramifications for the social acceptance of RET than then the typical risk-averse view assumes. Social conflict is a form of participation and societal assessment, in which a heterogeneity of actors and issues is involved. To impose analytical or administrative boundaries upon the phenomenon of social conflict would be both naive and incorrect, and may cause descriptive and normative problems. In policy practices, it may lead to addressing the wrong problem (Dunn, 2018; Hisschemöller & Hoppe, 2018), because the problem definition is not likely to match with that of other actors (e.g. citizens). Such a mismatch may eventually lead to backfiring (Wolf, 2016), for example, stalling of decision-making or more conflict. Avoiding conflicts can most basically be seen as a waste of energy: firstly, they may always arise, but given the contingency of controversy spillovers, one cannot predict when, where and how; secondly, there is something to be learned from conflict, which is an opportunity that deserves to be exploited. In this concluding section, we will reflect, firstly, on the implications for empirical research on social conflict as societal assessment, secondly, on the normative aspects of our account of social conflict, and thirdly, on the political nature of doing research on social conflict.

Let us start with implications for empirical research on social conflict as societal assessment. We argued that the process of societal assessment evolves as a rhizome: it never comes to an end. This makes it rather challenging to empirically approximate societal assessment. In fact, no method will be able to pinpoint the preference of the public-at-large. Instead, there are only methods that provide a snapshot or that make ‘proxies’ of what society wants, desires or accepts. The repertoire of methods to study societal assessment through social conflicts seems to be too limited, and not fitting the intrinsically temporal and dynamic -rhizomatic- nature of social conflict. We have to search for methods and frameworks that fit with this nature. Although social theories and methods are not well-equipped to understand processes of change (Pesch, 2015), theories and methods exist that can be used to understand how a social conflict around a particular planned energy infrastructure evolves, such as process theory (Langley, 1999). Event sequence analysis (Abbott, 1995; Boons et al., 2011), which builds on process theory, may be useful, as it allows the analytical breakdown of a wide range of temporally related events. It would also be beneficial to use methods that allow the identification of expectations and orientations of actors. We have, for instance, worked with Q-methodology, which allows for systematic empirical categorisation of ideas, images and beliefs. Such categorisations are much more fine-grained and empirically reliant than conventional labels about the motivations and orientations of actors (Cuppen et al., 2010; van de Grift et al., 2020).

In addition to expanding the methodological repertoire, the objects of observation in empirical research need to be expanded. We already mentioned the need to dive more deeply into powerful, yet understudied actors in social conflicts: project developers, energy companies, governmental bodies and all actors that engage with citizens and/or local communities in energy infrastructure planning ('studying up': Nader, 1972). Organisational theory and organisational ethnographies may provide fruitful avenues for understanding how these actors create the conditions and context in which conflicts emerge. Furthermore, the observation of controversy spillovers, as well as the spillover between different types of public engagement, demands a shift in attention towards *trajectories* of engagement and mechanisms of spillover between multiple social conflicts, rather than a focus on specific local RET projects or planning decisions.

The second point pertains to the normative view on the distinction between public and private interests in social conflict, and in public engagement more broadly. This distinction is key in policy and academic discourses on social acceptance. An example is the NIMBY discourse, which implies that private interests—as opposed to public interests—are considered an illegitimate basis for rejecting RETs. Also, the recurrent reference to Habermas' ideal speech situation in participatory literature is an example of how private interests are excluded as a justifiable starting point for discussing matters of public interest.

Habermas' work is concerned with how citizens can contribute to the establishment of a societal assessment. This aligns with the view on democracy as a collective endeavour in which the members of society come to discuss how to relate to those issues that are perceived as collective. For Habermas, the ideal speech situation is a 'counterfactual' condition, meaning that it is a condition that needs to be presupposed, even though it may not be fulfilled, in order to allow for a democratic dialogue (Habermas, 2014). The absence of power, one of the criteria of an ideal speech situation, will be unattainable in real life, as actors will differ in those aspects that constitute power, such as authoritativeness, knowledge, persuasiveness, and so on. This counterfactual criterion will never be fulfilled, but figures as guidance for organising and evaluating such deliberative processes.

Another counterfactual criterion introduced by Habermas is that it is not justified to derive a societal assessment from private concerns and interests (Pesch, 2019). This criterion opposes neo-liberal inspired approaches to societal assessment which only account for private orientations, with that ignoring the essence of societal assessment as an intrinsically public phenomenon. Excluding private orientations completely may not be desirable, let alone possible, in real life. Firstly, because companies that have commercial interests play a major role in RET development and as such cannot be discarded. The energy system, although there are institutional differences between countries, by nature is a system that involves public ***and*** private interests, for example, through consumer choice and private firms. Secondly, people may feel triggered to join a deliberative process because they are affected by a future development through concerns that are predominantly of a private nature. These private concerns may give rise to public orientations once actors become engaged in wider processes of societal assessment (Pesch et al., 2017a), as is illustrated by the earlier example of the NLVOW.

Indeed, seeing social conflict as a rhizome implies that there is no stringent demarcation between the public and private orientation of actors. This does not mean that the boundary between public and private is 'blurred', but that this cannot be assumed to be predetermined or static. In fact, actors may go back and forth between a private and a public orientation. For instance, a social conflict can follow from the way in which people are affected in their personal life sphere, when they are confronted by a planned RET project in their neighbourhood. Subsequently, members from a community may mobilise themselves and form a group that develops a public standpoint on that RET in general (cf. Pesch et al., 2017b).

Our third reflection concerns the political nature of our work as researchers on social conflict. We may become actors in social conflicts when we are studying them. Also, our funding opportunities are more and more framed by the agendas of industries and government who adhere to instrumental, managerial views on social conflict and expect us to help them in reducing conflict or 'managing' citizens. Not only are we, as researchers, sometimes close to the actors that are involved in social conflicts, scholarly findings also tend to become part of the expectations and conceptions maintained by actors in conflicts. We have started this account by criticising the conventional picture of public acceptance and social conflict, not just because we consider it to be incorrect, but because it influences the process of societal assessment in a negative way due to its rather simplistic starting points. As researchers, we have to be aware that our work motivates certain expectations and assumptions—we are part of the rhizome, and as such have to account for the conceptions and categorisations that we introduce.

Acknowledgements This chapter was written based on research done within three projects funded by the Responsible Innovation programme by the Netherlands Organisation for Scientific Research (NWO), project numbers 313-99-303, 313-99-322, MVI.16.007.

Note

1. This illustration is based on personal communication with actors involved in the NLVOW and NPBO.

References

Abbott, A. (1995). Sequence analysis: New methods for old ideas. *Annual Review of Sociology, 21*(1), 93–113. https://doi.org/10.1146/annurev.soc.21.1.93

Aitken, M. (2009). Wind power planning controversies and the construction of 'expert' and 'lay' knowledges. *Science as Culture, 18*(1), 47–64.

Aitken, M. (2010). Why we still don't understand the social aspects of wind power: A critique of key assumptions within the literature. *Energy Policy*. https://doi.org/10.1016/j.enpol.2009.11.060

Anderson, C. (2013). The networked minority: How a small group prevailed in a local windfarm conflict. *Energy Policy, 58*, 97–108. https://doi.org/10.1016/j.enpol.2013.02.048

Barnett, J., Burningham, K., Walker, G., & Cass, N. (2012). Imagined publics and engagement around renewable energy technologies in the UK. *Public Understanding of Science, 21*(1), 36–50. https://doi.org/10.1177/0963662510365663

Barry, J., & Ellis, G. (2011). Beyond consensus? Agonism, republicanism and a low carbon future. In P. Devine-Wright (Ed.), *Renewable energy and the public: From NIMBY to participation* (pp. 29–42). Earthscan. https://doi.org/10.4324/9781849776707

Batel, S. (2018). A critical discussion of research on the social acceptance of renewable energy generation and associated infrastructures and an agenda for the future. *Journal of Environmental Policy and Planning, 20*, 356–369. https://doi.org/10.1080/1523908X.2017.1417120

Batel, S. (2020). Research on the social acceptance of renewable energy technologies: Past, present and future. *Energy Research & Social Science, 68*. https://doi.org/10.1016/j.erss.2020.101544

Bell, D., Gray, T., & Haggett, C. (2005). The 'social gap' in wind farm siting decisions: Explanations and policy responses. *Environmental Politics, 14*(4), 460–477. https://doi.org/10.1080/09644010500175833

Bogner, A. (2012). The paradox of participation experiments. *Science, Technology & Human Values, 37*(5), 506–527. https://doi.org/10.1177/0162243911430398

Boons, F., Spekkink, W., & Mouzakitis, Y. (2011). The dynamics of industrial symbiosis: A proposal for a conceptual framework based upon a comprehensive literature review. *Journal of Cleaner Production, 19*(9–10), 905–911.

Burningham, K., Barnett, J., & Walker, G. (2015). An array of deficits: Unpacking NIMBY discourses in wind energy developers' conceptualizations of their local opponents. *Society & Natural Resources, 28*(3), 246–260.

Chilvers, J., Pallett, H., & Hargreaves, T. (2018). Ecologies of participation in socio-technical change: The case of energy system transitions. *Energy Research and Social Science, 42*, 199–210. https://doi.org/10.1016/j.erss.2018.03.020

Cotton, M., & Devine-Wright, P. (2012). Making electricity networks "visible": Industry actor representations of "publics" and public engagement in infrastructure planning. *Public Understanding of Science, 21*(1), 17–35.

Cuppen, E. (2018). The value of social conflicts. Critiquing invited participation in energy projects. *Energy Research & Social Science, 38*, 28–32. https://doi.org/10.1016/J.ERSS.2018.01.016

Cuppen, E., Breukers, S., Hisschemöller, M., & Bergsma, E. (2010). Q methodology to select participants for a stakeholder dialogue on energy options from biomass in the Netherlands. *Ecological Economics, 69*(3), 579–591. https://doi.org/10.1016/j.ecolecon.2009.09.005

Cuppen, E., Ejderyan, O., Pesch, U., Spruit, S., van de Grift, E., Correljé, A., & Taebi, B. (2020). When controversies cascade: Analysing the dynamics of public engagement and conflict in the Netherlands and Switzerland through "controversy spillover". *Energy Research and Social Science*. https://doi.org/10.1016/j.erss.2020.101593

Deleuze, G., & Guattari, F. (1987). *A thousand plateaus*. University of Minnesota Press.

Devine-Wright, P. (2011). Public engagement with large-scale renewable energy technologies: Breaking the cycle of NIMBYism. *Wiley Interdisciplinary Reviews: Climate Change, 2*(1), 19–26.

Devine-Wright, P. (2013). Explaining "NIMBY" objections to a power line: The role of personal, place attachment and project-related factors. *Environment and Behavior, 45*(6), 761–781. https://doi.org/10.1177/0013916512440435

Dunn, W. N. (2018). Using the method of context validation to mitigate type III errors in environmental policy analysis. In M. Hisschemöller, R. Hoppe, W. N. Dunn, & J. R. Ravetz (Eds.), *Knowledge, power, and participation in environmental policy analysis* (pp. 417–436). Transaction Publishers. https://doi.org/10.4324/9781351325721-25

Fiorino, D. J. (1990). Citizen participation and environmental risk: A survey of institutional mechanisms. *Science, Technology & Human Values, 15*(2), 226–243.

Gross, C. (2007). Community perspectives of wind energy in Australia: The application of a justice and community fairness framework to increase social acceptance. *Energy Policy, 35*(5), 2727–2736. https://doi.org/10.1016/j.enpol.2006.12.013

Groves, C. (2010). The futures of causality: Hans Jonas and Gilles Deleuze. *Causality and Motivation*. https://doi.org/10.1515/9783110329575.151

Habermas, J. (1970). Towards a theory of communicative competence. *Inquiry, 13*, 363–372.

Habermas, J. (2014). *Truth and justification*. Wiley & Sons.

Hisschemöller, M., & Cuppen, E. (2015). Participatory assessment: Tools for empowering, learning and legitimating? *The Tools of Policy Formulation: Actors, Capacities, Venues and Effects, 2001*, 33–51. https://doi.org/10.4337/9781783477043.00013

Hisschemöller, M., & Hoppe, R. (2018). Coping with intractable controversies: The case for problem structuring in policy design and analysis. In M. Hisschemöller, R. Hoppe, W. N. Dunn, & J. R. Ravetz (Eds.), *Knowledge, power, and participation in environmental policy analysis* (pp. 47–72). Transaction Publishers. https://doi.org/10.4324/9781351325721-4

Langley, A. (1999). From for theorizing strategies. *Academy of Management Review, 24*(4), 691–710.

Leach, M., & Scoones, I. (2007). Mobilising citizens: Social movements and the politics of knowledge. *IDS Working Paper*, January, 37. http://w.drc-citizenship.org/system/assets/1052734544/original/1052734544-leach_etal.2007-mobilising.pdf

Leeuwis, C. (2000). Reconceptualizing participation for sustainable rural development: Towards a negotiation approach. *Development and Change, 31*(5), 931–959. https://doi.org/10.1111/1467-7660.00184

Mansbridge, J., Bohman, J., Chambers, S., Christiano, T., Fung, A., Parkinson, J., Thompson, D. F., & Warren, M. E. (2012). A systemic approach to deliberative democracy. In *Deliberative systems: Deliberative democracy at the large scale* (pp. 1–26). Cambridge University Press. https://doi.org/10.1017/CBO9781139178914.002

Marris, C. (2015). The construction of imaginaries of the public as a threat to synthetic biology. *Science as Culture, 24*(1), 83–98. https://doi.org/10.1080/09505431.2014.986320

Nader, L. (1972). *Up the anthropologist. Perspectives gained from studying up*. In D. Hymes (Ed.), *Reinventing anthropology* (pp. 284–311). Pantheon B.

Pesch, U. (2015). Tracing discursive space: Agency and change in sustainability transitions. *Technological Forecasting and Social Change, 90*, 379–388. https://doi.org/10.1016/j.techfore.2014.05.009

Pesch, U. (2019). Elusive publics in energy projects: The politics of localness and energy democracy. *Energy Research & Social Science, 56*, 101225.

Pesch, U., Correljé, A., Cuppen, E., & Taebi, B. (2017a). Energy justice and controversies: Formal and informal assessment in energy projects. *Energy Policy, 109*. https://doi.org/10.1016/j.enpol.2017.06.040

Pesch, U., Correljé, A., Cuppen, E., Taebi, B., & Vvan de Grift, E. (2017b). Formal and informal assessment of energy technologies. In L. Asveld, R. Van Dam-Mieras, T. Swierstra, S. Lavrijssen, K. Linse, & J. van den Hoven (Eds.), *Responsible innovation 3* (pp. 131–148). Springer.

Pesch, U., Huijts, N., Bombaerts, G., Doorn, N., & Hunka, A. (2020). Creating "local publics": Responsibility and involvement in decision-making on technologies with local impacts. *Science and Engineering Ethics*, 1–20.

Pesch, U., Spekkink, W., & Quist, J. (2019). Local sustainability initiatives: Innovation and civic engagement in societal experiments. *European Planning Studies, 27*(2), 300–317. https://doi.org/10.1080/09654313.2018.1464549

Reed, M. S., Vella, S., Challies, E., de Vente, J., Frewer, L., Hohenwallner-Ries, D., Huber, T., Neumann, R. K., Oughton, E. A., Sidoli del Ceno, J., & van Delden, H. (2018). A theory of participation: What makes stakeholder and public engagement in environmental management work? *Restoration Ecology, 26*, S7–S17. https://doi.org/10.1111/rec.12541

Renn, O., Blättel-Mink, B., & Kastenholz, H. (1997). Discursive methods in environmental decision making. *Business Strategy and the Environment, 6*(4), 218–231. https://doi.org/10.1002/(SICI)1099-0836(199709)6:4<218::AID-BSE117>3.0.CO;2-G

Renn, O., & Schweizer, P. J. (2009). Inclusive risk governance: Concepts and application to environmental policy making. *Environmental Policy and Governance*. https://doi.org/10.1002/eet.507

Stirling, A. (2008). "Opening up" and "closing down": Power, participation, and pluralism in the social appraisal of technology. *Science, Technology & Human Values, 33*, 262–294.

Susskind, L., & Field, P. (1996). *Dealing with an angry public: The mutual gains approach to resolving disputes*. Simon and Schuster.

Taylor, C. (2002). Modern social imaginaries. *Public Culture*. https://doi.org/10.1215/08992363-14-1-91

Usher, M. (2013). Defending and transcending local identity through environmental discourse. *Environmental Politics, 22*(5), 811–831. https://doi.org/10.1080/09644016.2013.765685

van de Grift, E., Cuppen, E., & Spruit, S. (2020). Co-creation, control or compliance? How Dutch community engagement professionals view their work. *Energy Research and Social Science, 60*, 101323. https://doi.org/10.1016/j.erss.2019.101323

van der Voort, N., & Vanclay, F. (2015). Social impacts of earthquakes caused by gas extraction in the Province of Groningen, The Netherlands. *Environmental Impact Assessment Review, 50*, 1–15. https://doi.org/10.1016/j.eiar.2014.08.008

Verloo, N. (2018). Social-spatial narrative: A framework to analyze the democratic opportunity of conflict. *Political Geography, 62*, 137–148. https://doi.org/10.1016/j.polgeo.2017.11.001

Walker, G., & Cass, N. (2007). Carbon reduction, 'the public' and renewable energy. *Area, 39*(4), 458–469. https://doi.org/10.1111/j.1475-4762.2007.00772.x

Walker, G., Cass, N., Burningham, K., & Barnett, J. (2010). Renewable energy and sociotechnical change: Imagined subjectivities of "the public" and their implications. *Environment and Planning A, 42*, 931–947.

Walker, G., Devine-Wright, P., Barnett, J., Burningham, K., Cass, N., Devine-Wright, H., Speller, G., Barton, J., Evans, B., Heath, Y., Infield, D., Parks, J., & Theobald, K. (2011). Symmetries, expectations, dynamics and contexts: A framework for understanding public engagement with renewable energy projects. In P. Devine-Wright (Ed.), *Renewable energy and the public: From NIMBY to participation* (pp. 1–14). Earthscan.

Walsh, B., van der Plank, S., & Behrens, P. (2017). The effect of community consultation on perceptions of a proposed mine: A case study from southeast Australia. *Resources Policy, 51*, 163–171. https://doi.org/10.1016/j.resourpol.2016.12.006

Welsh, I., & Wynne, B. (2013). Science, scientism and imaginaries of publics in the UK: Passive objects, incipient threats. *Science as Culture, 22*(4), 540–566. https://doi.org/10.1080/14636778.2013.764072

Wolf, E. E. A. (2016). *Conflict reconsidered? The boomerang effect of depoliticization.* July, 5–8.

Wolsink, M. (2000). Wind power and the NIMBY-myth: Institutional capacity and the limited significance of public support. *Renewable Energy, 21*(1), 49–64. https://doi.org/10.1016/S0960-1481(99)00130-5

Wolsink, M. (2019). Social acceptance, lost objects, and obsession with the 'public'—The pressing need for enhanced conceptual and methodological rigor. *Energy Research and Social Science, 48*, 269–176. https://doi.org/10.1016/j.erss.2018.12.006

Wüstenhagen, R., Wolsink, M., & Bürer, M. J. (2007). Social acceptance of renewable energy innovation: An introduction to the concept. *Energy Policy, 35*(5), 2683–2691. https://doi.org/10.1016/j.enpol.2006.12.001

Wynne, B. (2001). Creating public alienation: Expert cultures of risk and ethics on GMOs. *Science as Culture, 10*(4), 445–481. https://doi.org/10.1080/09505430120093586

Wynne, B. (2006). Public engagement as a means of restoring public trust in science—Hitting the notes, but missing the music? *Public Health Genomics, 9*(3), 2110220. https://doi.org/10.1159/000092659

Wynne, B. (2007). Public participation in science and technology: Performing and obscuring a political–conceptual category mistake. *East Asian Science, Technology and Society: An International Journal, 1*(1), 99–110.

PART VI

Overcoming Localism and Spatial Determinism

CHAPTER 10

Provincial Polyphasia: Community Energy Generation and the Politics of Sustainability Transition in Alberta, Canada

Mike Gismondi and Lorelei Hanson

Introduction

Broad public acceptance of energy transition in Canada's oil rich province of Alberta is complicated (Hanson & Kahane, 2018). While Alberta's participation in the global move to renewables and decentralised energy is evident, mention of any significant move away from oil and gas, particularly oil sands expansion, is seen as anti-Albertan, a threat to economic livelihoods and an existential risk to the province (Hislop, 2020). Our work lies at the centre of this contradiction, examining the actors, arguments and contextual forces that constrain or compel acceptance of community energy generation and energy transition in Alberta.

Context-specific interventions in particular places provide opportunities to connect theoretical problems and practical solutions. We examine renewable energy development—particularly community-developed solar

M. Gismondi (✉) • L. Hanson
Athabasca University, Athabasca, AB, Canada
e-mail: mikeg@athabascau.ca; lorelei.hanson@athabascau.ca

S. Batel, D. Rudolph (eds.), *A critical approach to the social acceptance of renewable energy infrastructures*,
https://doi.org/10.1007/978-3-030-73699-6_10

projects—in Alberta. Community energy (CE) is defined as "electricity and/or heat production on a small, local scale that may be governed by or for local people or otherwise capable of providing them with direct beneficial outcomes" (Walker et al., 2012, p. 194). We are attracted to CE's potential for broadening ownership of renewables (IRENA, 2018), but also as a "potential springboard to local collective environmental action beyond energy" (Berka et al., 2020).

Alberta is an interesting and complicated setting to undertake analysis of CE. Foremost, Alberta is home to the third largest proven oil reserves in the world (Government of Alberta, 2019), over three quarters of which are bitumen deposits (also called oil or tar sands) that many contend are one of the world's dirtiest fuel sources (Davidson & Gismondi, 2011; Efstathiou & Orland, 2018). Beyond this, fossil fuel extraction dominates Alberta's economy, politics and culture. Social acceptance of CE in this context is best understood as shaped by socio-economic, cultural and structural conditions. As a number of social acceptance researchers contend (Batel, 2018; Batel & Devine-Wright, 2015; Cuppen, 2018), social conflicts about energy projects are often complex, and generally not only about individual attitudes and local issues but also entwined with broader regional, national and international contexts that shape peoples' knowledge, representation of issues and practices (Batel, 2012).

Fossil fuels are key to Alberta's electricity production. About 91% of electricity in Alberta is produced from them: 43% from coal, 49% from natural gas (CER, 2019). As a result, Alberta's electricity sector produces more greenhouse gas (GHG) emissions than any other province because of its size and reliance on coal-fired generation (Piper & Green, 2017). In 2016, Alberta's GHG emissions were three times the national average (Natural Resources Canada, 2018). Individual use of electricity, though, represents barely 18% of all electricity consumption. Industry consumes well over 50% and even more fossil fuel fired electricity is generated and consumed onsite behind commercial metres. Weis et al. construe that "[t]he dominance of the industrial sector… has resulted in its disproportionate influence on electricity policy and planning" (2016, p. 519) in Alberta.

While reliant on the oil and gas industry for jobs and social funding, Alberta is home to some of the highest quality and most accessible wind resources (Pembina, n.d.), best solar resources in Canada (SESA, 2018) and Albertans lead the country in energy efficient net-zero housing (Park, 2019). Yet, just over 9% of electricity is generated by renewables such as wind, hydro and biomass. Photovoltaics barely register in the provincial breakdown at less than 1% (CER, 2019). At the time of writing, the adoption of solar generation has increased, but is largely corporate owned utility-scale developments.

Attitudinal acceptance of renewables is also evident. A national poll found over 58% of Albertans consider climate change an emergency currently or in the future, and 75% Albertans believe we should shift to clean and renewable energy sources (Abacus Data, 2019). A researcher of Albertans' attitudes about energy describes two concurrent rationalities operating: "transitioning to renewables is seen as a different thing than transitioning away from fossil fuels, specifically transitioning away from oil and gas. The idea was that renewables are nice to have, but the fundamentals of our economy aren't going to change" (Hislop, 2020). Social psychologists describe the "coexistence of competing and even contradictory meanings, not only within the same culture and groups, but also within the same individual" as cognitive polyphasia (Batel et al. 2016, p. 736). Like Batel et al. however, we maintain that individuals become "aware of different meanings insofar as they are part of communities/groups where different representations are available" (p. 741) and shaped by context. Holding competing and contradictory views over what direction to take in an environmental turning point can be either "a mechanism to support or resist change" (Batel, 2012, p. 1). The researcher's role is to explore "if and how different meanings are used and in which contexts" (Batel et al., 2016, p. 736). This approach fits with our reading of CE in Alberta, a provincial context that is "(at least…) double-edged—it is what is there beforehand and it is what social agents (can) make out of it—depending on the power relations defining that context" (Batel 2012, p. 6).

To examine provincial polyphasia we focus on solar CE actors trying to disrupt Alberta's partially deregulated and privatised electricity regime. Data was collected from 2015 to 2019 and includes 26 semi-structured interviews, and notes from meetings and collaborations with CE actors in Alberta. The interviews focused on specific encounters of solar CE actors advancing their projects, and general challenges for distributive community generation to gain a foothold in Alberta. We interviewed niche actors including farmers, activists and citizens, and intermediaries who are retailers, municipal employees, co-op advocates, engineers and market analysts. Informed consent was obtained from all participants and interviews conducted in accordance with considerations of participant confidentiality, researcher transparency and collaborative learning.[1]

Theorising Social Acceptance and Power

To surface and disentangle the complexities of social acceptance of CE from our interviews, we combined strands of social acceptance research with scholarship focused on critical studies of power in systems change (Avelino & Wittmayer, 2016; Boonstra, 2016; Brisbois, 2019). Brisbois (2019), for example, identifies three overlapping dimensions of power struggles: (1) instrumental or overt power (corporate manipulation, influence on state power, financial sway); (2) subtle structural ways incumbent actors use to control or block innovators, favour established interests, set agendas, and shape policy preferences, and (3) dominant discourses and narratives that privilege certain transition paths over others. We're curious about how Alberta CE actors confront each dimension of power, and whether power is shifting. Each approach enters the issue differently, but "conceives of actors, networks, innovations and changes as mutually defining" (Haxeltine et al., 2017, p. 70), in which conflicting roles and struggles over meanings support or inhibit acceptance of renewables (Wüstenhagen et al., 2007).

Building on Avelino et al. (2016, pp. 560–561) we want to avoid a simple David and Goliath interpretation (Strachan et al., 2015) and acknowledge three cross cutting dimensions of power in transition politics. One, how struggles over "power in transition processes is not concentrated at a particular level (e.g., 'niche' or 'regime') or within specific actors, but [instead] different dimensions of power are dispersed across interrelated agents at numerous levels" (Avelino et al., 2016, p. 560). Two, how material artefacts or the power of things (e.g., the materiality of the electricity grid) shape individual and institutional behaviour and choices. Three, how place making is political as actors construct, re-imagine and struggle over historical and spatial context. As with critical social acceptance theory, struggles over meaning-making and re-presentation are actions, not just word games (Batel et al., 2016). Individuals and groups acquire critical knowledge as they foster innovations. They begin to see regulatory and technical constraints as social constructs, and use this tacit knowledge to question established assumptions, conventions, and ways of seeing or to fashion counter-metaphors, imaginaries, or narratives (Smith et al., 2016). Equally, by rallying the emotions of place many have begun to re-imagine Alberta's sunny and windy landscapes as places of possibility to initiate transition to renewables (van Veelen & Haggett, 2017). This is not an easy task when the historical industrial landscape of pump-jacks, tailings ponds, and flare gas is normalised and emotionally charged with a positive aura to be defended, even when located in the middle of grain fields, forests, or suburbs.

Our story begins with Alberta's regional and resource history, and its adoption of a deregulated electricity system.

Situating the Growth of Renewables in Alberta

In the first half of the twentieth century, Alberta's settlers and governments were subordinated in part to eastern Canadian bankers and politicians who controlled immigration, land, loans, public investment and natural resources in the western provinces. Hard times gave rise to resentment, as well as a progressive cooperative agrarian populism with roots in Alberta's rural electrification associations, cooperative farm and fuel supply, local credit unions, and a provincial politics opposing the "plutocratic economic parasites of Central Canada" (Epp, 2006, p. 742). With the discovery of oil in the province in 1947, Alberta went from a struggling agricultural economy to a prosperous province. Cooperativism waned and by 1972, a newly elected right of centre Conservative party began to develop a world-class petrochemical industry creating a state supported economic regionalism, described by Richards and Pratt (1979) as 'prairie capitalism.' Alberta's economy flourished. Oil prices rose with the global oil crisis, but Westerners' hopes of resource sovereignty were crushed when in 1980 the federal government used its powers to create the National Energy Program that restricted Alberta's oil exports and kept prices low to support eastern Canadian industries and consumers. Alberta's oil boom went bust: thousands of people lost companies, jobs and homes and new grievances against Ottawa joined old ones.

In the early 1990s, a new Conservative premier, Ralph Klein, applied neoliberal principles, economic rules and structures on Albertans to boost recovery. A new narrative—the 'Alberta Advantage'—legitimated privatising government lands and services, gutting social supports, expanding the oil sands, and embracing free market language and ideology (Taft, 1997). Alberta neoliberals considered business as the solution to poverty, social inequality and even environmental protection, rather than government policy and programmes (Filax, 2008). Their brand of neoliberalism combined trust in the market with distrust of the federal government and a strong sense that an Alberta solution is always better than anything that comes from central Canada (Bregha, 2006). Alberta's transition to a deregulated electricity system started here and was complete by 2003.

In 2015, there was a surprising landslide election victory by the left-leaning New Democrats, who retained power for only four years. In that time, they introduced a carbon tax, an energy efficiency programme and renewable energy incentives. Their 2015 Climate Leadership Plan included a commitment over five years to reinvest $645 million of the carbon levy revenue into energy efficiency and renewable CE programmes to reduce greenhouse gases and create green jobs (Government of Alberta, 2016). The Plan resulted in a reduction of 3.4 million tonnes of GHG emissions and $330 million in energy savings (Energy Efficiency Alberta, 2018). They also passed the Renewable Electricity Act. To reach 30% renewable generation by 2030, they began to phase out 6300 megawatts of coal-fired electricity production, two-thirds to be replaced by renewables (BLG, 2017). They provided residential and commercial incentives, promised $200 million to launch a Community Generation Plan, supported by a Small-scale Generation Regulation (Government of Alberta, 2018), and conducted auctions for renewable electricity, resulting in about 1350 MW of promised utility-scale renewable electricity development (Government of Alberta, 2019). Just before they were ousted, the New Democrats funded a Community Generation Capacity Building Program to support early development of projects but CE remained quite limited. In April 2019, the far-right United Conservative Party assumed power and immediately rescinded the carbon levy, ending many of these renewable programmes.

In spite of interest in renewables, weighty ideas of the importance of the fossil fuel economy continue to saturate Alberta's political air and culture, animating federal/provincial tensions and informing the politics of electricity system changes. As one of our co-op informants explains, it's

> *a mentality of Alberta. The land of plenty abundance. We've got oil and gas reserves that are growing every year, or least according to our reserves auditors, but we just really haven't gotten down to maybe a sense of need for renewable energy because we've got the cheap and easy stuff kind of coming out of all orifices of Alberta.* (I-19)

Encountering Dispositional Power: Alberta's Electricity Grid

Unlike most provinces in Canada, the Alberta government has never owned and operated a utility company. From the mid-1950s until 2000, Alberta's electricity was provided by private and a few municipally owned utilities. Power plants were constructed and operated under government direction, and electricity was distributed for cost plus a profit-margin of 9% to 14% (AESO, 2016). In 1996, Alberta began restructuring its electricity system away from traditional regulation towards a more market-based system. Alberta is geographically 20% larger than France, with 4.37 million residents, 26,000 kms of transmission lines and an installed electricity generation of 16,106 MW. The majority of power produced and consumed within Alberta now flows through a real-time energy market power pool, in which the revenue for generators, as well as the cost for consumers, is based on a competitive price determined by supply and demand on the market. The power pool is managed by the Alberta Electric System Operator (AESO), an arms-length association of the Alberta government. The price for electricity fluctuates from hour to hour influenced by short term outages at generation facilities and extreme weather (which can raise the price as high as $1,000/MWh) and by surplus events when too much power is produced (which can push the price as low as $0/MWh). Transmission, though, is regulated in Alberta under a cost-of-service model where customers pay for the full costs of operating the system, plus a reasonable rate of return. The transmission grid is largely owned by four public, for-profit companies. The AESO plans system additions to serve new load and generation, operates the system in real-time, and determines the charge for using the transmission system (AUC, 2019). The distribution systems throughout the province are largely owned and operated by private companies, some municipalities, and a few co-ops, who are all called distribution facility operators (DFOs). The DFOs plan upgrades and enhancements to the electric distribution system. And finally, electricity retailers purchase bulk power at the pool price from the wholesale market based on their customers' projected demand. While the management and operation structure of Alberta's electrical grid is fairly unique in North America, its utility structure is like most others, and this allows solar generators to connect to feeder lines typically connected to the high-voltage distribution grids at substations, without necessarily needing long transmission lines (Nadkarni & Hastings-Simon, 2017).

Although encouraged by government policy support, many of the CE actors that we spoke with felt stymied as they engaged electrical system actors in Alberta to seek grid connectivity. One Alberta Solar Cooperative (ASC) member described his initial confrontation with the electrical systems experts working for the major corporations and regulatory boards:

> *When technical people have the information… they have an incredibly disproportionate amount of power. They can simply say, 'Well, keep dreaming. That doesn't work.' Or, 'Do you want to give a billion dollars to AESO to buy a new software package to make that happen?' You know, and they make you sound stupid…. That's where the power dynamic is the rawest, I suppose.* (I-10)

As gatekeepers of the grid, experts use their authority to limit CE, representing grassroots leaders as deficient, yet declaring support for renewables in principle. Dispositional and instrumental power dynamics operate against CE in structural ways too. The government handles distributed CE generators as if they are large utility-scale developments, overcomplicating the approval process. Small CE generators (1 to 5 MW) must comply with multiple pieces of legislation, conduct wildlife studies that span a full year, and undertake extensive public consultation to ensure acceptance by their site neighbours. To achieve compliance, even in the early stages, studies can cost hundreds of thousands of dollars (I-5). CE proponents must also compete with corporations for limited connections to the distribution substations and grid access. Local farmer and ASC board member Jordan Webber explains how this functions in his rural Alberta county:

> It is more cost effective for smaller electricity generators to connect to a distribution substation instead of a transmission connection in order to deliver electricity to the grid. However, each substation can only take up a limited amount of capacity depending on its rating and the local grid…. (Todd, 2017)

Small community generators like ASC can barely afford to purchase one substation queue position. However, according to Webber, large corporations gamble and purchase multiple substation connection options:

> There are two substations in Starland County and there are already three applications on each substation with the Alberta Utilities Commission to connect projects…To make matters worse, the system favours larger

companies with capital because a 50 per cent deposit is required to hold a place in the queue. But CE projects don't have that kind of capital at such an early stage. As a result, it is difficult for community projects to compete with the large industry incumbents. (Todd, 2017)

Competition leaves even less room for CE start-ups to get a foothold as large integrated renewable energy companies, many European, can secure project sites with nearby grid access more easily. As one of the Starland County co-op advocates explains:

> *The vultures are circling for sure from outside… several large international companies including a large French one, which is the largest solar installer in the world…[are]ready with significant financial backing to come in and scoop up any capacity right? …They've got… billions of dollars behind them and so that's a pretty big, pretty big Goliath that we've got to work with. So there's the outside investors who've got… the solar capacity and… they see a market here.* (I-7)

A Starland farmer who also does economic development work with the county identified that solar companies outside Alberta (i.e., Ontario, Germany, France, etc.) have secured leases, targeting land near substations (I-11):

> *…their reach is significant. Like, it's hard to find a farm that has not been canvassed to do a lease for solar or wind or something… my farm has been asked by three different companies if they can put installations on our land…. [T]hey're paying 10 times market rent: just to put that in perspective, we pay $50 an acre to rent farmland for farmland. You know, these guys are taking a full section, 260 hectares or 640 acres, which is a big piece of land, and they're signing a lease on that for $600 to a thousand dollars an acre.* (I-11)

At the level of the province, another energy co-op advocate finds the Alberta government's electricity system "*is dominated by powerful corporate interests that are making things difficult for little entrants*" (I-10). Large multinational corporations including ATCO, ENMAX, EPCOR and AltaLink own the transmission lines and four major distribution companies—EPCOR, ENMAX, ATCO and FortisAlberta—dominate distribution. According to this co-op advocate, this means CE actors entering Alberta's electricity system are not, as the hegemonic discourse of Alberta advantage claims, joining a free market, but rather an "*oligopoly*" (I-10).

Even AESO, who is mandated by provincial legislation to act in the public interest, is managed by former private sector actors, many of whom put corporate interests first as this informant explains:

> *There are a lot of employees of AESO… that ensure fair competition and so forth, but most of the employees have formerly worked for one of the big players, whether they're from ATCO or ENMAX or… You know, the very people that are running the electricity system in Alberta… and this is a third level of barrier: the infrastructure.* (I-10)

Over the past fifteen years, verified instances of corruption and the government passing legislation against public interest and in favour of corporate interests have been ignored (Nikiforuk, 2011). Employee attitudes in roles within the current electricity system can also present barriers to CE actors trying to advance projects (I-8, I-9). "Institutions are inhabited" by people (Hallett & Ventresca, 2006) and their mindsets often reinforce existing ways the system functions. One CE intermediary informed us that:

> *…where it is necessary to interact with the grid, you have no choice but to engage with their processes and their policies and their tariffs and their costs… [T]hose pieces of the system… provide a level of inertia… I am not going to refer to them as an obstacle but they have a lot more baggage that they are carrying… I mean that they have done things in a certain way for a lot of years and they are filled with humans that are accustomed to a utility world where they know what is best for the consumer… And all those people who work inside those organisations, or many of them, are facing dramatic disruption to the roles that they play, the expectations placed on them. They are having to make adjustments to their worldview. And humans find that kind of change difficult. And so it isn't going perfectly smoothly yet.* (I-9)

Finally, the twenty-year old experiment with deregulation of the electricity sector, acts like a common-sense hegemonic discourse and continues to elicit the allegiance of many system actors. As one interviewee mused:

> *You know, there are some bureaucrats who I've talked to and… they talk consistently of the integrity of the system. There's a lot of push by the incumbent large generators and particularly distribution companies on keeping the system the same even though it disadvantages Alberta ratepayers… some of the bureaucrats that are there were involved in deregulation and trying to make it work. And so they… want to keep that system whole to some extent even though it's got*

> *some significant gaps... So you're dealing with... the heritage system around that and you got to get their minds changed around... particularly distribution companies which have a sweet deal within the current system and would want to resist any kind of significant change to that.* (I-7)

The Technical and Material Challenges Faced by CE Actors

Alberta's electricity system creates a number of technical and material challenges and obstacles for CE projects. In particular, transmission and distribution line operators and owners have insider knowledge. They influence and control the functioning and layout of the electrical grid with respect to substations, switches, energy loads and how the electricity system components and its regulatory structures fit together. The asymmetry of power between CE actors and systems experts means the latter not only influence the character of the technological system but also interpret how the system works. Their influence is embodied in the technology itself (Klein & Kleinman, 2002, p. 33). As is often the case, finding and interpreting 'publicly available' information requires a level of expertise most system outsiders do not possess. Yet, frequently this information is key to determining the viability of a community generation project. According to one of our co-op advocates, the easiest way to stall CE distributed generation is for wire owners to say:

> *'That won't work.' And what they mean by that is, simply a quick technical assessment of what you're asking for... to the end of, you know, you can't do that in the current system. We'd have to buy very expensive software to give you those numbers, we'd have to install substations that let the power go both ways, we'd have to put up all kinds of new lines, and on and on and on, right?... They own the infrastructure, so it's very hard, then, to get that proprietary information, to be able to challenge that very statement that says, 'It won't work.'* (I-10)

A solar field development implemented by the Métis Nation of Alberta (MNA) demonstrates well the many entanglements of material and social power imbalances, and how having someone on your side who understands the intricacies of the operation of the electricity grid system can make or break your project. Located in central Alberta, Métis Crossing is Alberta's first major Métis cultural interpretive site. As part of the MNA's climate change deliberations, they are in the process of developing a solar generation field on their 512-acres comprising traditional Métis settler's river lots. The project will provide renewable electricity to a new interpretive centre and sell electricity to the nearby Town of Smoky Lake. Initially the MNA were planning to build a 7 to 8 MW solar field but after consulting with their engineers, they learnt that "*there are good reasons to stay below 5 MW. A project is 'non-dispatchable' under 5 megawatts… so AESO can't make you shut off*" (I-5). In other words, when there is surplus power in Alberta's electricity grid, AESO can force any distributed generator of 5 MW or larger to shut down to help balance electricity supply and demand. Of course, shutting down generation facilities a number of times per year has financial implications for community generators like the MNA (they would receive zero revenue) and can weaken business plans.

The MNA was also able to avoid a large upfront capital cost because their engineers understood technical aspects of the grid. When the MNA originally contacted the wire owners for their area,—ATCO—about grid connectivity, ATCO informed them that they would need to link to a substation about 50 kilometres away. Building distribution lines is a major expense and ATCO's proposal would have made their project financially unfeasible. However, the MNA's consultants identified a closer switch on the feeder line that could be used, which required only 5 kilometres of line upgrades at a cost of $36,000, saving them over $200,000 in capital start-up costs, and making the project viable again (I-5).

Regulatory Politics

Another example illustrates how grid regulators use meaning-making to re-present technology in ways that thwart distributed energy generation. An ongoing review by the Alberta Utilities Commission, which adjudicates transmission applications, penalties and system rules, (AESO, 2016), is examining a recent application by AESO to amend their tariffs. "*There was a significant change proposed to the independent system operator tariff that would have a big impact on how distributed generators are compensated for load on the local distribution grid, in terms of avoided wires costs*" (I-9). As one of the electrical engineering intermediaries supporting the CE sector explained, "*for certain generators, [the credits are] up to 25% of potential revenue*" (I-9). Many CE projects will be hard-pressed to proceed without the transmission bonus, let alone generate a surplus for community benefits (I-5). A coalition of CE advocates hired Joe Peters, an independent grid connectivity specialist, to provide an alternative 'reading' of the mechanical and technical evidence of distributed generation impacts on feeder lines, substations and grid performance. According to one of the ASC members, Peters counter-acted assertions by AESO about negative technical impacts from CE, offering instead evidence of its positive merits. In Peters' words:

> … located much closer to loads, ultimately requiring less transmission infrastructure to provide the same levels of energy supply and reliability. This creates the opportunity to better leverage existing transmission and distribution infrastructure, generally reducing the cost of the overall system. (2019, p. 3)

Peters' submission challenged additional efforts by AESO to charge distributed energy generators for use of substations on the distribution lines, including costs for recent and even future upgrades. The proposed charges, called substation fractioning, could potentially cost millions of dollars. Peters argued that small distributed solar electricity generators only use a slight portion of the local substation, by-passing transformers connected to centralised transmission feeds, and reducing pressures to enlarge substations. The changes proposed by AESO, Peters concluded, "can be expected to act as a deterrent to the development of DCG (decentralised community generation) throughout Alberta" (2019, p. 4). According to the Canadian Solar Industry Association (CanSIA), AESO's application amounted to changing rules in the middle of the game:

> The AESO's proposal to fundamentally change the world in which DG participants live, and have been living since the early 2000s when DG credits were approved and found to be in the public interest, is a significant and fundamental change… and… will impact project economics for new DG in the future. (2018)

CanSIA also revealed that while AESO failed to mention this in their submission, AESO's predecessors established transmission credits to reward gas companies for capturing waste flare gas and producing electricity to power gas plants distributed across Alberta. CanSIA wondered why transmission credits considered 'just' for Alberta's industrial generators, should, after all those years, be considered 'unjust' subsidies for community distributed generation, if in the end both improve the overall system.

In late 2019, the Alberta Utilities Commission ruled in favour of AESO but after much protest from the renewables sector, the decision will be reviewed. It's a case of materiality twisted up with politics. Not surprising, as the CEO of the rural electricity cooperative EQUS concludes, "*it has not been easy for a generator to connect to the utility systems, because they do not want you there*" (I-4). One of the intermediary CE actors confirmed, "*There's big money in energy*" (I-9) and the incumbent system players are interested in maintaining their share of the pie. Even our informant who works for a utility declared, "*there's a huge gap between the present power grid dominated by centralised production and creating an electricity grid with electricity generated and consumed locally*" (I-3).

Conclusion

We examined the conflicting values of Albertans concerning social acceptance for CE in a province where cultural identity, employment, livelihood and politics is heavily dominated by the oil and gas industry. Our analysis brought two interdisciplinary fields of inquiry into tension: social acceptance scholarship and theories of power in sustainability transitions to extend understandings of: (1) how conflicting values of groups and individuals are shaped by larger institutional and socio-historical contexts; (2) the struggles CE actors navigate over meaning-making across the electricity system; and (3) the broader material and institutional struggles encountered by CE actors. Using these critical theoretical lenses helped us analyse macro and micro sticking points faced by CE actors in a Canadian province divided over the future of its oil industry. We discovered everything from corporate competition and unwieldy government approval processes, to active resistance by private wire companies and regulatory and state actors who undermine newcomer confidence, re-interpret technology systems, re-define and re-present rules, remove financial incentives, or simply say no.

We highlight how the social re-presentation of CE in Alberta—whether questions of the dispositional power of grid owners and corporate competitors, the inertia of electrical system bureaucrats, differing cultural and discursive representations of decentralised generation, or the asymmetries of power between citizens and experts as each interprets and manoeuvres regulations or technical aspects of the grid, including who has the authority to make change—constitute a provincial polyphasia. Our research highlights the many kinds of power encountered by CE actors at various levels and forms, and how that is reflected in struggles over the social representation of transition and social acceptance of or resistance to renewable energy. Within Alberta, conflicting values operate to constrain and restrain an energy transition and yet, concurrently offer the possibility for intervention and innovative grassroots change. The CE actors we interviewed are not ushering in transformative social change or energy transition in Alberta. They represent instances of changing social relationships at two and sometimes, three dimensions of the system that could be built upon. For example, the ASC, who draw on a cooperative ethos, and the MNA who draw on shared Indigenous culture and entrepreneurial identity, both see themselves as piloting a model of community solar generation larger than their own interests, a prototype that can be scaled out to support community sustainability across the province. Too, they see themselves springing from a long line of prairie entrepreneurs and innovators ready to shake up the current system with an alternative distributed energy nodes in a province of centralised, corporatized energy politics and influence.

Note

1. All quotes from community actors interviewed are shown in italics and distinguished from each other by a number (e.g., I5) to protect the informant's anonymity.

References

Abacus Data. (2019, Summer). Climate emergency polling. https://abacusdata.ca/wp-content/uploads/2019/08/Climate-Emergency-Polling-July-2019-RELEASE.pdf

AESO—Alberta Electrical System Operator. (2016). Guide to understanding AB's electricity market. https://www.aeso.ca/aeso/training/guide-to-understanding-albertas-electricity-market/

AUC—Alberta Utilities Commission. (2019). History of the electricity market. *Electric Industry.* http://www.auc.ab.ca/pages/history-electric-industry.aspx

Avelino, F., Grin, J., Pel, B., & Jhagroe, S. (2016). The politics of sustainability transitions. *Journal of Environmental Policy & Planning, 18*(5), 557–567.

Avelino, F., & Wittmayer, J. M. (2016). Shifting power relations in sustainability transitions: A multi-actor perspective. *Journal of Environmental Policy & Planning, 18*(5), 628–649.

Batel, S. (2012). Commentary on re-presenting (and) cognitive polyphasia. *Papers on Social Representations, 21*(1), 10–15.

Batel, S. (2018). A critical discussion of research on the social acceptance of renewable energy generation and associated infrastructures and an agenda for the future. *Journal of Environmental Policy & Planning, 20*(3), 356–369.

Batel, S., Castro, P., Devine-Wright, P., & Howarth, C. (2016). Developing a critical agenda to understand pro-environmental actions: Contributions from social representations and social practices theories. *Wiley Interdisciplinary Reviews: Climate Change, 7*, 727–745.

Batel, S., & Devine-Wright, P. (2015). Towards a better understanding of people's responses to renewable energy technologies: Insights from social representation theory. *Public Understanding of Science, 24*(3), 311–325.

Berka, A. L., MacArthur, J. L., & Gonnelli, C. (2020). Explaining inclusivity in energy transitions: Local and community energy in Aotearoa New Zealand. *Environmental Innovation and Societal Transitions, 34*, 165–182.

BLG—Borden, Ladner & Gervais. (2017, April). New opportunities in the Alberta electricity market. https://blg.com/en/News-And-Publications/new-opportunities-in-the-alberta-electricity-market

Boonstra, W. (2016). Conceptualizing power to study social-ecological interactions. *Ecology and Society, 21*(1), 21. https://doi.org/10.5751/ES-07966-210121

Bregha, F. (2006). National energy program. *Canadian Encyclopedia*. http://www.thecanadianencyclopedia.com/index.cfm

Brisbois, M. C. (2019). Powershifts: A framework for assessing the growing impact of decentralized ownership of energy transitions on political decision-making. *Energy Research & Social Science, 50*, 151–161.

CanSia to AUC. (2018, September 5). Re: Alberta Electric System Operator ("AESO") ISO tariff application to Alberta Utilities Commission, proceeding 2294.

CER—Canada Energy Regulator. (2019). Provincial and territorial energy profiles—Alberta. https://www.neb-one.gc.ca/nrg/ntgrtd/mrkt/nrgsstmprfls/ab-eng.html

Cuppen, E. (2018). The value of social conflicts: Critiquing invited participation in energy projects. *Energy Research & Social Science, 38*, 28–32.

Davidson, D. J., & Gismondi, M. (2011). *Challenging legitimacy at the precipice of energy calamity*. Springer Press.

Efstathiou, Jr., J. & Orland, K. (2018, March 26). Alberta, home to one of the world's dirtiest oil sources, wants to be green. *BNN Bloomberg News*. https://www.bnnbloomberg.ca/one-of-world-s-dirtiest-oil-sources-wants-to-be-green-but-how-1.1037805

Energy Efficiency Alberta. (2018). 2017–2018 annual report. https://efficiencyalberta.ca/itsbeenabigyear/pdf/EEA_2017_Annual_Report.pdf#page=3

Epp, R. (2006). Two Albertas: Rural and urban trajectories. In M. Payne, D. Wetherell, & C. Cavanaugh (Eds.), *Alberta formed, Alberta transformed* (pp. 726–746). University of Alberta Press.

Filax, G. (2008). Race-ing Alberta-ness. *The Ardent*, 50.

Government of Alberta. (2016). Energy efficiency and community energy in Alberta: Discussion document. https://www.alberta.ca/documents/EEAP-DiscussionDocument.pdf

Government of Alberta. (2018). Industry profiles 2018—mining and oil and gas extraction industry. https://work.alberta.ca/documents/industry-profile-mining-oil-and-gas-extraction.pdf

Government of Alberta. (2019). Oil sands facts and statistics. https://www.alberta.ca/oil-sands-facts-and-statistics.aspx

Hallett, T., & Ventresca, M. J. (2006). Inhabited institutions: Social interactions and organizational forms in Gouldner's *Patterns of industrial bureaucracy. Theory and Society, 35*(2), 213–236.

Hanson, L. L., & Kahane, D. (2018). Introduction. In L. L. Hanson (Ed.), *Public deliberation on climate change: Lessons from Alberta climate dialogue* (pp. 33–66). Athabasca University Press.

Haxeltine, A., Pel, B., Wittmayer, J., Dumitru, A., Kemp, R., & Avelino, F. (2017). Building a middle-range theory of transformative social innovation: Theoretical pitfalls and methodological responses. *European Public & Social Innovation Review, 2*(1), 59–77.

Hislop, M. (2020, January 30). What Albertans really think about energy transition. Interview with Professor Melanie Thomas. https://energi.media/energy-transition-climate-series/energy-narratives/you-may-be-surprised-what-albertans-really-think-about-the-energy-transition-interview-with-prof-melanee-thomas/

IRENA. (2018). Coalition for action. Community energy: Broadening the ownership of renewables. http://coalition.irena.org/-/media/Files/IRENA/Coalition-for-Action/Publication/Coalition-for-Action_Community-Energy_2018.pdf

Klein, H. K., & Kleinman, D. L. (2002). The social construction of technology: Structural considerations. *Science, Technology, & Human Values, 27*(1), 28–52.

Nadkarni, K., & Hastings-Simon, A. (2017). *Alberta community solar guide: Organizing and owning community solar PV projects*. Pembina Institute.

Natural Resources Canada. (2018, November 2). Energy and greenhouse gas emissions (GHGs). https://www.nrcan.gc.ca/energy/facts/energy-ghgs/20063

Nikiforuk, A. (2011, Feb. 8). Scandal pugs into the oil sands. The Tyee. https://thetyee.ca/News/2011/02/08/AlbertaElectricity/

Park, D. (2019, October 31). Canada's largest net zero community in the centre of Edmonton. *Juggling Dynamite*. https://jugglingdynamite.com/2019/10/31/canadas-largest-net-zero-community-in-the-center-of-edmonton/

Pembina Institute. (n.d.). Community-owned renewables-fact sheet. https://www.pembina.org/reports/community-owned-re-fact-sheet.pdf

Peters, J. (2019, January 15). Peters energy solutions. Written evidence. AUC Proceeding 22942, AESO 2018 ISO Tariff application.

Piper, L., & Green, H. (2017). A province powered by coal: The renaissance of coal mining in late twentieth-century Alberta. *Canadian Historical Review, 98*(3), 532–567.

Richards, J., & Pratt, L. (1979). *Prairie capitalism: Power and influence in the new West.* McClelland and Stewart.

SESA-Solar Energy Society of Alberta. (2018). https://solaralberta.ca

Smith, A., Hargreaves, T., Hielscher, S., Martiskainen, M., & Seyfang, G. (2016). Making the most of community energies: Three perspectives on grassroots innovation. *Environment and Planning A, 48*(2), 407–432.

Strachan, P. A., Cowell, R., Ellis, G., Sherry-Brennan, F., & Toke, D. (2015). Promoting community renewable energy in a corporate energy world. *Sustainable Development, 23*(2), 96–109.

Taft, K. (1997). *Shredding the public interest: Ralph Klein and twenty-five years of one-party government.* University of Alberta Press.

Todd, L. (2017, September 6). Alberta's community energy story: The economic benefits of renewable energy. *Pembina Institute blog.* https://www.pembina.org/blog/alberta-s-community-energy-story

van Veelen, B., & Haggett, C. (2017). Uncommon ground: The role of different place attachments in explaining community renewable energy projects. *Sociologia Ruralis, 57*, 533–554.

Walker, G., Simcock, N., & Smith, S. J. (2012). Community energy systems. In S. J. Smith (Ed.), *International encyclopedia of housing and home* (1st ed., pp. 194–198). Elsevier.

Weis, T., Thibault, B., & Miller, B. (2016). Alberta's electricity future. In L. Adkin (Ed.), *First world petro-politics: The political ecology and governance of Alberta* (pp. 499–526). University of Toronto Press.

Wüstenhagen, R., Wolsink, M., & Bürer, M. J. (2007). Social acceptance of renewable energy innovation: An introduction to the concept. *Energy Policy, 35*(5), 2683–2691.

CHAPTER 11

People-Place Bonds, Rhetorical Meaning-Making and "Doing Acceptance" to a Renewable Energy Infrastructure: Postcolonial Insights from the Global South

Rafaella Lenoir-Improta and Andrés Di Masso

Introduction

In order to understand the social acceptance of renewable energy technologies (henceforth, RET), several studies have grounded their explanations on the role played by a number of individual psychological constructs (e.g. perceptions, beliefs, attitudes, behaviours). However, psychological research on social responses to RET has paid little attention to the political and cultural contexts that frame the psychological experiences that bond people to places (for exceptions, see e.g. Aitken et al., 2008; Batel & Devine-Wright, 2017; Cowell et al., 2011; Devine-Wright, 2009; Farias, 2017). In this chapter, we argue that a rigorous study of citizens' evaluative responses to RET must necessarily consider the culturally shared and

R. Lenoir-Improta (✉) • A. Di Masso
University of Barcelona, Barcelona, Spain
e-mail: rafaellalenoir@uam.es; adimasso@ub.edu

S. Batel, D. Rudolph (eds.), *A critical approach to the social acceptance of renewable energy infrastructures*,
https://doi.org/10.1007/978-3-030-73699-6_11

politically embedded frames of meaning that shape those attitudes, especially when they channel historically normalized representations of place and people-place relations that derive from colonial practices.

At a local level, the implementation of RET tends to have an impact on the subjective bonds that people living in the affected areas have established with their surroundings (Brownlee et al., 2015; Devine-Wright, 2009; Devine-Wright & Howes, 2010). Subjective people-place relations may have political connotations when they have an argumentative value to contest, reject, and delegitimize the implementation of RET. For instance, residents may report having strong feelings of place attachment and place-identifications that appear to be threatened by the construction of a new wind farm, thereby warranting, by implication, that this environmental proposal is wrong and should be reconsidered (e.g. Batel et al., 2015; Devine-Wright & Howes, 2010; Zografos & Martínez-Alier, 2009; Späth, 2012). Conversely, place-based identity processes and place attachments can also be contextually functional to legitimize the construction of wind farms. In this case, the latter could be represented as preserving the historical continuity of the cultural heritage of the place (Lenoir-Improta, 2017), or could be associated with ideas of progress (i.e. new jobs, reinforcing the local economy), and working as a symbol of contemporaneity (Lenoir-Improta & Pinheiro, 2011). In both cases, the expression of the psychological implications of building RET operates as a political statement grounded in culturally shared notions about the meaningful ways that connect people to "their" places.

At a macro-structural level, place identity can catalyse hegemonic power relations between the Global North and the Global South (Batel & Devine-Wright, 2017, Dixon & Durrheim, 2000 — outside of the RET context). The subordinator-subordinate relationship is reproduced through energy policies, particularly through the implementation of RET–a phenomenon labelled 'energy colonialism' (Batel & Devine-Wright, 2017). In this context, "advanced" knowledge is exported to "weaker" countries, thus perpetuating the current, but at the same time, historically continuous global power relations. In this postcolonial frame, northern technologies are accepted without much controversy (and if there is any, it is swiftly neutralized).

In consequence, it is not uncommon that RET implementation programmes carry the basis of neoliberal capitalist logics, characterized by fostering an apolitical or acritical position that draws people's attention away from the hegemonic relations in which energy interventions are grounded (Batel & Devine-Wright, 2017; Swyngedouw, 2010; Walker, 2009). Ostensibly, for instance, in the name of climate change mitigation or protecting the world's energy reserves, renewable technologies devised by multinationals generate a series of environmental injustices (Howe, 2014). The latter can be suffered by local communities when they do not directly benefit from that energy due to the centralized nature of the electrical network. Social disempowerment and social disarticulation may provoke that local communities are not aware of the possibility of fighting back the exploitation of their resources, whose main beneficiaries (in terms of consumption and of economic gains) are foreign people. Furthermore, changing policies to introduce new initiatives may occur before social rights are guaranteed, thereby increasing social injustices. Therefore, while local communities are not aware of their social rights to benefit from compensations, or they do not have an adequate representation in negotiations and decision-making processes, structural inequalities will tend to be prolonged.

The perpetuation of hegemonic (postcolonial) power relations and lack of social knowledge about social rights can be illustrated, for example, by drawing on Brazil's first wind farm (Parque Eólico de Rio do Fogo), built and managed by a multinational company. The nearest neighbour was located very close to the wind turbines (less than 1 km away), inhabitants have a close relationship with the land, and the turbines were placed in front of their houses. Nevertheless, their streets lacked public lighting, although some inhabitants hoped that this issue would be resolved as a positive outcome of the new wind farm. Furthermore, residents were not financially compensated in any way. It was striking that, in this context, in a previous study the majority of the residents expressed a positive opinion of the new wind farm (Lenoir-Improta & Pinheiro, 2011).

Exploring Postcolonial RET Acceptance in the Global South

Critical analyses of the neoliberal policies involved in the implementation of RET have already demonstrated the mechanisms behind the rhetoric of environmental injustices and energy colonialism in Global North-Global South relations (Batel & Devine-Wright, 2017; Finley-Brook et al., 2008; Groves, 2015). However, only a few studies have explored this from a Global South perspective. Some exceptions are the analysis of carbon trading interpreted as a global process of domination and perpetuation of power relations (Bumpus & Liverman, 2008), violating human rights in traditional local communities in the construction of renewable energies projects that can be considered as a form of "carbon colonialism" and "green authoritarianism" (Farias, 2017; Finley-Brook & Thomas, 2011; Howe, 2014).

To address this gap, and reinforcing existing studies in the Global-South perspective, this chapter illustrates the technological North-South colonization, drawing on some results of a study on the role of people-place bonds in articulating and legitimizing the implementation of the first wind farm in the far South of Brazil. The focus was twofold. First, we analysed the discursive construction of the meanings of the wind farm and residents' relationships with the territory, as displayed in the narratives of the locals and the media. The aim was to explore the contextually situated, socially constructed, and negotiated nature of these frames of evaluative meanings. Second, we explored to what extent energy colonialism shapes this territory's cultural imaginary and its environmental transformations, which are channelled and reproduced through evaluative discourses about RET. Specifically, the questions that we sought to answer in this study were: What symbolic elements of the local culture and the political structure are stressed when unfolding discourses of acceptance or rejection of the wind farm? How do place-talk and arguments construct specific person-place psychological bonds that contribute to the perpetuation and legitimization of energy colonialism?

We used a psycho-discursive perspective to conduct a detailed exploration of the cultural and political dimensions of the impact of RET on the processes of social acceptance and rejection. This form of analysis involves pinpointing the discursive strategies through which people construct, negotiate, and challenge the meaning of their psychological relationships with their environment, and the meaning of place itself, interpreting if and how these meaningful constructions are strategically used to "warrant and legitimize, or to challenge and undermine, specific arrangements of social relations which are power-driven and are relatively central to the social order" (Di Masso et al., 2017, p. 95).

Context

This study was conducted in Cerro Chato, a rural district of the city of Santana do Livramento, where the Cerro Chato Wind Farm (henceforth CCWF) was built. CCWF is composed of 45 wind turbines constructed among 21 cattle farms. The wind farm construction began in 2010 and started working in 2012. Cerro Chato is close to the border with Uruguay. Due to its location, this area has been marked by intense struggles regarding its territorial delimitation. Residents have a strong identity-based and cultural connection with *gaucho* traditions. One of the main characteristics of this culture is its self-image as warrior people, defenders of their lands and traditions (Chelotti & Pessôa, 2006). It must be noted that, nowadays, *gaucho* culture has a political connotation. Following Haesbaert (1988) and Kaiser (1999), the declining middle class has tended to capitalize on regional culture, which is relatively well-established in the area, in an attempt to homogenize the complex and emergent urban middle-class and to reassure the old oligarchy, reinvigorated by its cultural roots in its struggle to maintain its monopoly of the countryside.

In addition, and somewhat contradictorily, the identity of this region is constructed around less positive meanings. The economy of Santana do Livramento is based on agriculture and livestock farming. Between the 1930s and the 1970s, the city experienced intense economic development in connection with the livestock farming industry (Ferreira, 1959). After this period, a deep crisis hit the region, triggering an economic recession in the industrial and agricultural sectors due to competition with areas closer to the great Brazilian capitals (e.g. Sao Paulo and Rio de Janeiro). This economic recession started a process of decline that remains until today. At the time of the data collection for this study (from 2011 to 2013), Santana do Livramento was among the ten Brazilian cities with the highest level of population decline (IBGE, 2010) due to the exodus to more economically prosperous parts of the country. As a result of this process, the city comes to be self-recognized as "the city that had but no longer has".

Like Santana do Livramento, the Cerro Chato district had also undergone a period of intense development at the height of the city's economic boom, thanks to its livestock production. Until 1970s, Cerro Chato had small markets, a school, and traditional celebrations. Nowadays, there are no public services and only a handful of livestock farmers live in the region.

Due to the fact that it was the first wind farm in the far south of Brazil, CCWF is a new reality for the Santana do Livramento city. Its construction offered job for the residents, but only during the construction. The municipality won royalties for public taxes. The farm-owners win a percentage (2%) in the energy production of each wind turbine in their lands.

Sample and Analysis

The study started examining all the news ($n = 230$) about the construction of the wind farm published in the most locally read newspaper, the regional "A Plateia". We selected all news where CCWF appears as the main topic, from the first news published (2009) to the last news published in the year that CCWF started to produce energy (2012). The aim was to explore the processes of representation of the wind farm. We focused on how evaluative positions related to the wind farm construction were based on place-meanings and place identity formulations. News was examined using thematic analysis (Braun & Clarke, 2006) and argumentative analysis (Billig, 1987).

On the other hand, we analysed the discourses of 25 interviewees who spent their daily lives in Cerro Chato and directly experienced changes in the area and in their lives attributed to the wind farm. Interviews have been conducted during the phase of wind farm construction. The analytic focus was set on how the interviewees experienced the novelty of having a wind farm near their homes, and how their relationships with the place could be relevant (or not) to articulate an attitudinal position of acceptance or rejection. In theoretical terms, we assumed that by examining the discursive strategies that variably make and unmake versions of place, of people-place bonds, and of evaluative positions towards environmental changes, we can unveil the micro-politics of environmental issues and energy policies (see Di Masso et al., 2014). The extracts of press analysis and interviews presented bellow illustrate (1) how contradictory identity elements are used to legitimize the construction of CCWF, and (2) the process of integration of the CCWF into the local identity.

Recovering the Gaucho Image: Cultural Place Identity, Local Empowerment and Self-Esteem

As described in the previous section, *gaucho* identity is historically grounded in a series of territorial disputes that shaped the geography and the geopolitics of this area. *Gauchos* are the romanticized main characters of many stories about the struggles to defend and establish the territory. Nowadays, *gauchos* are still regarded as representatives of this ideal: brave people who take care of and defend what is theirs. Therefore, in this region, identity is cemented by the territory. One's pride to be oneself is the pride of being a *gaucho*, and to be a *gaucho* is to belong to and defend the land that historically defines one's identity. Thus, identity-based pride remains and is reinforced by the presence of *gauchos* in the place. In contrast, unsurprisingly, the economic crises of recent decades have led to some degree of "identity depression" in the region.

The fragment presented below reveals how the meaning of the wind farm is constructed upon the basis of both cultural elements: the positive identity of being a *gaucho* and the confusion associated with being unable to find a way out of the crisis devastating the region. These two contradictory elements also constitute the identity of the region.

Extract 11.1

The people of Santana have yet to realize how valuable this investment is. In terms of size and importance, not money. But since most people know nothing about engineering, they don't realize what is being built. (…) The people from Santana who have not visited Cerro Chato have no idea what work is being done there. It's a transformation. (…) It is only natural for them to raise their eyebrows to express their confusion and surprise. But that is a temporary thing. You just need to see it to be convinced. More than the [economic] investment, which is obviously important, there will be an enduring physical investiment; a moral investment, which will help recover the self-esteem of Santana – you can see this in the look of fascination in Paixão Côrtes' face upon arriving in Cerro Chato after decades without setting foot in the region. Well defined by a Uruguayan reporter, yesterday he 'looked like a *guri*, he was transfixed'. ("Aprendendo com o vento" [Learning with the Wind], 20/01/2011).

In this fragment, the recovery of the city's self-esteem and empowerment is linked to the construction of the wind farm. This argumentative construction indirectly conveys a negative social representation of the local population. By characterizing it as a way of restoring the city's self-esteem, the wind farm argumentatively emerges as a socially relevant element for a community defined by its close relationship with the territory (hence the importance of the "physical investment").

Interestingly, *gaucho* identity is illustrated through the figure of Paixão Côrtes, a well-known local artist born in Cerro Chato and who was one of the main icons of the movement that sought to extol *gaucho* traditions in the 1960s. The use of expressions that reflect regional identity, such as "transfixed *guri*" (a gaucho's colloquial expression to refer to "boy") further reinforces the wind farm's positive contribution to local identity. Thus, the wind farm latches on to historical-cultural heritage as a figure that also represents local identity, acquiring reliability and admiration. Therefore, the CCWF's meaning is blended with the identity construction of the gauchos, a brave and warrior people, working as a symbolic foothold to increase community self-esteem and local empowerment.

From a rhetorical perspective, it must be noted here that this construction of the wind farm as resonating with the symbolic imaginary of the gaucho, actually neutralizes potential resistances against this infrastructure, as it would entail undermining place identity itself. Furthermore, this extract is an allusion to the typically colonial epistemic subordination of popular knowledge to technical knowledge (Barnett et al., 2012; Bermann, 2001; Bumpus & Liverman, 2008; Cotton & Devine-Wright, 2012; de Sousa Santos, 2010b), in this case about energy and its advantages ("those who know nothing about engineering"). This lack of knowledge about the scientific dimensions of RET is also a discursive 'locus' that perpetuates a form of political power that undermines criticism on the grounds of a technical lack of expertise, preventing active social participation in debates and decision-making process.

The Rhetoric of Progress: The Wind Farm Bridging *Gaucho* Traditions and the Modern World

News about the wind farm frequently referred to it as a sign of a "new moment", a "new era" for the city. The decadent present vanishes, paving the way for a promising future in which the wind farm is a protagonist. The following extracts from newspaper and interviews illustrate this view:

Extract 11.2

Our Sant'Ana do Livramento needed to leave its mark on this modern world, which is full of diversity and has a variety of technological advances that are often surprising and follow one another very quickly. We obviously have potentials that characterize us economically, socially, and culturally, but it is clear that we needed something to really buttress this municipality known for its *gaucho* hospitality, its water, the ideal genetics of its bovine and ovine cattle, and its cult of traditions. ("Eólica: uma marca para Santana do Livramento" [Wind Power: A Brand for Santana do Livramento], 21/12/2009)

Extract 11.3

In the fields near the border, <u>along with our traditional products</u> (ovine and bovine cattle), there stand tall <u>those giants of modernity and technology</u>, the Wind Turbines. ("Livramento conquista mais 5 parques" [Livramento Conquest 5 More Wind Farms], 21/08/2011)

Extract 11.4

What do they think? That we're going to make a fortune with each turbine. I know we aren't. (...) But the impression they have, the outsiders, is: lucky the ones that were here. I believe that they think exactly how (...) I feel like someone who won the lottery. The only thing is that I don't have is the money [laughs] just the feeling. (...) When I see my bank account, it's the same. So, I didn't get the money, but it's incredible to see that. It's incredible to see these things because <u>this is an act of progress</u> right there, wide-open, in front you. (Interview. 21)

Again, these extracts reveal a link between the wind farm and elements that define the economic and cultural traditions of the region. However, unlike traditional elements, the wind farm is associated not only with economic salvation but is also regarded as an identity-based symbol that looks to the future: a path to modernity that will lead the region to progress.

In Extracts 11.2, 11.3, 11.4, the CCWF is associated with somewhat contradictory traits. Local traditions are used to connect the wind farm with elements that confer identity to the place. This association presents the wind farm as a bridge linking tradition with modernity. This process causes the city to take on a new meaning that differs from all other parts of its identity, which had been established a priori. Unlike prior identity elements, which are traditional and tend to be regarded as old-fashioned or passé, the wind farm is modern and innovative.

Nevertheless, the wind farm has another, less evident meaning connected to energy-based colonization (Batel & Devine-Wright, 2017). These extracts demonstrate the postcolonial perpetuation of the North-South imaginary of progress, with foreign technology being regarded as contemporaneity, as progress, and therefore representing the only way to tackle the city's stagnation. This notion of progress connected with the innovations provided by a temporally advanced Global North (which "arrived earlier" at technological solutions and now "shares them" with the South) exemplifies what de Sousa Santos (2010a) has labelled the 'monoculture of linear time'. That is, true and legitimate knowledge (which is, at the same time, the one produced according to Western-North scientific standards and taken for granted as a universal system of truths) can only take one direction: that which involves "moving forward" according to the scientific findings and narratives typically developed in the white, European, and English-speaking world. In other words, knowledge, development, innovations, and globalization are characteristics of "developed" countries, which represent contemporaneity and stand in contrast with everything else (i.e. "underdevelopment"). At the same time, the traditional is regarded as old-fashioned. In consequence, this logic produces non-existence, labelling as 'backwards' everything that, according to the temporal norm, is asymmetric relative to what is labelled as 'advanced'; therefore, western modernity has produced the non-contemporaneity of the contemporary. Those who do not heed the edicts of the great world powers, even if they are part of contemporaneity, undergo a process of "residualization" (de Sousa Santos, 2010a), which causes them to be branded as obsolete, primitive, or traditional. In this context, the wind farm plays a role that *gaucho* cultural and economic traditions could not have adopted in this "modern world".

In this regard, Batel (2018) proposes analysing the relationship between RET and local social needs. The CCWF has covered longstanding historical needs, both identity-related and economic. The wind farm has empowered the city and its population, although maybe only symbolically and temporarily. Beyond the economic and infrastructure needs covered by the project, something else is added to the discourse of acceptance: the message that the city has returned to the contemporary world, thus getting closer to "first world" urban centres. As we referred to in Extract

11.2, the wind farm seems to symbolically replace the spirit of the traditional local activity, recovering their lost place in the contemporary world.

Social and Political Implications for a Critical Agenda

So far, we have argued that cultural elements that constitute a territory's place identity and symbolism are employed by the interviewees and the most read newspaper to articulate assessments of the wind farm that capitalize on the postcolonial imaginary, thereby legitimizing the economic interests of Global North actors in the region. To do so, place-related discourses anchor the colonial ideology of modernity (progress, technique, science, economic profit) in the symbolic imaginary of the gaucho as the epitome of local tradition, thereby endorsing social acceptance. Regarding the cultural elements deployed in assessments of the CCWF, we observed that both the media and the local population employed positive (*gaucho* tradition) and negative (decadence and stagnation of the city) people-place bond rhetoric to legitimize the construction of the wind farm. This analysis has at least two broad implications for a critical agenda oriented to ensure effective sociopolitical changes: (1) at social democratization level and (2) to energy decolonization.

Social Democratization

The process of RET implementation can be a resource to promote social participation, not only in a concrete project but also in other debates on the future of the city, thus contributing to fairer decision-making in both social and environmental terms and enabling it to become an instrument for social democratization (Aitken et al., 2008; Batel, 2018; Cowell et al., 2011). However, for a critical participation in decision-making processes to occur, and for RET to become a critical instrument for social democratization, it is necessary for the people to understand the energy sector and public policy. The lack of social dissemination of knowledge about energy issues is often concealed by the difficulty of understanding this technical-scientific information (Bermann, 2001; Wüstenhagen et al., 2007). Furthermore, this reason is also used to justify the limited information dissemination and social involvement in debates on energy policy. Nonetheless, exercising social participation in subjects related to energy policy, can in and of itself help to construct a more attentive and informed society, one that is able to take an active role in public debates on energy generation and use. In other words, laws should not be imposed on the population. Society should play an active role in the debate and the decision-making processes surrounding energy sector issues (Bell et al., 2005; Bermann, 2001; Cotton & Devine-Wright, 2012; Cowell et al., 2011; Zografos and Martínez-Alier, 2009).

However, 'energy ignorance' may be strategic for the implementation of governmental measures regarding energy. Insufficient social knowledge about topics related to energy management generates subjection, thus facilitating social manipulation. In this regard, social exclusion operates as 'manoeuvrable mass' for imposing electrical projects (Devine-Wright, 2005). Likewise, the interests of private institutions masquerading as State bodies (Pol et al., 2006) may also underlie the measures adopted in the energy sector. This means that energy ignorance also becomes strategic for these measures to be implemented (e.g. electrical sector privatizations, benefits for certain companies when proposing new projects) (Bermann, 2001; Scheer, 2009).

Energy Decolonization

Critical social involvement in debates on energy will not in and of itself guarantee democratization and energy justice. Full knowledge of energy innovations must include awareness of the overall energy situation, including national and local energy policies as well as the global logics of energy mobility in a Global North-to-South direction, from "developed" to "underdeveloped" countries.

According to Quijano (2000) and de Sousa Santos (2010a), in order to stay in the contemporary world the imaginary of North-South progress promotes the need to submit to foreign technologies, because they are better than those developed locally. At the same time, the Southern Hemisphere's imaginary regards elements coming from the outside as modern and cutting-edge, thus perpetuating the colonizer-colonized relationship. In other words, in the name of energy security, distribution equality, and climate change mitigation, longstanding power relations between the Global North and South are perpetuated, with the South "de-existed" (de Sousa Santos, 2010b) by large multinationals and governments, but being indispensable to conduct experiments and sell new technologies. This creates an imperceptible argument in favour of the neoliberal manoeuvre of exporting (and colonizing) knowledge and technology, thus reproducing colonial power relations.

This perspective is particularly clear in renewable energy transitions. Typically, renewable energy policies and technologies put into practice around the world come from a global north "formula". Global South countries import renewable energy policies and technologies from the Global North. It is a common fact that the owners of the wind farms and their technologies come from European and North America countries. To the extent that Global North technologies are perceived by civil society as more "developed" (more modern, high-tech, updated) new wind farms are perceived as improving the quality of the city. In our study, when people described the advent of the wind farm as if they were "winning the lottery" understood as a metaphor of progress, and when people reported feeling that the wind farm was a landmark in the access to the contemporary world, they were giving testimonies, against the backdrop of a historically imposed and subjectively assumed state of "underdevelopment" [sic], of the ideological workings of energy colonialism.

Therefore, for true social democratization to be possible, it is necessary, on the one hand, to pursue energy decolonization, which must spring from social awareness of this veiled colonization. On the other hand, we should be able to contribute to render more visible and appreciate local knowledge, out of a dichotomic and epistemically hierarchical framework (Mouffe, 2013). In other words, energy infrastructure and policies have perpetuated and produced forms of inequity, vulnerability and injustices (Batel, 2018; Groves, 2015; Scheer, 2009; Walker, 2009). In consequence, for true environmental and energy justice to be achieved, it is necessary to consider the need for energy decolonization through the appreciation of local knowledge and accept the plurality of existence, which includes both local and external elements. Therefore, debates on energy and citizen involvement in decision-making should be grounded in awareness of these power relations between the Global North and South, leaving some space for exploring the potential of local knowledge. This consciousness should also be present in debates of government-level energy policies, by re-politicizing the debate on energy, through the encouragement of citizen participation in critical decision-making in the energy field, promoting both democratization and environmental justice (Batel, 2018; Cowell et al., 2011; Groves, 2015; Walker, 2012).

Final Considerations

This study contributes to the emerging literature on RET sensitive to the historically situated, cultural and political dimensions of people-place bonds, extending previous studies on the powerful role of place-discourses in the process of acceptance of RET and, simultaneously, in perpetuating historical and cultural power relations at local and macro level. Furthermore, we also pointed out the possibility of renewable energies to be an instrument for a more democratic and just society whether horizontal and critical citizen participation in the decision-making process take place, considering local knowledge and diversity, counterbalancing historical and cultural power relations from Global-North to Global-South, and adopting a plural perspective.

We encourage future research on energy development framed as a systemic function of Global North-South technological colonization. The perpetuation of Global North-South power relations and energy colonization through RET implementation, and assessments on the quality of energy policies, must especially take into account the level of "policy importation" from the Global-North. We also call for a critical exploration of the quality of social participation in the decision-making process.

References

Aitken, M., McDonald, S., & Strachan, P. (2008). Locating 'power' in wind power planning processes: The (not so) influential role of local objectors. *Journal of Environmental Planning and Management, 51*(6), 777–799.

Barnett, J., Burningham, K., Walker, G., & Cass, N. (2012). Imagined publics and engagement around renewable energy technologies in the UK. *Public Understanding of Science, 21*(1), 36–50.

Batel, S. (2018). A critical discussion of research on the social acceptance of renewable energy generation and associated infrastructures and an agenda for the future. *Journal of Environmental Policy & Planning, 20*(3), 356–369.

Batel, S., & Devine-Wright, P. (2017). Energy colonialism and the role of the global in the local responses to new energy infrastructures in the UK: A critical and exploratory empirical analysis. *Antipode, 49*(1), 3–22.

Batel, S., Devine-Wright, P., Wold, L., Egeland, H., Jacobsen, G., & Aas, O. (2015). The role of (de-)essentialisation within siting conflicts: An interdisciplinary approach. *Journal of Environmental Psychology, 4*, 149–159.

Bell, D., Gray, T., & Hagget, C. (2005). The 'social gap' in wind farm siting decisions: Explanations and policy responses. *Environmental Politics, 14*, 460–477.

Bermann, C. (2001). *Energia no Brasil: Para quê? Para quem?—Crise e alternativas para um país sustentável*. Livraria da Física.
Billig, M. (1987). *Arguing and thinking. A rhetorical approach to social psychology*. Cambridge University Press.
Braun, V., & Clarke, V. (2006). Using thematic analysis in psychology. *Qualitative Research in Psychology, 3*(2), 77–101.
Brownlee, M. T. J., Hallo, J. C., Jodice, L. W., Moore, D. D., Powell, R. B., & Wright, B. A. (2015). Place Attachment and marine recreationist' attitude toward offshore wind energy development. *Journal of Leisure Research, 47*(2), 263–284.
Bumpus, A., & Liverman, D. (2008). Accumulation by decarbonisation and the governance of carbon offsets. *Economic Geography, 84*(2), 127–155.
Chelotti, M. C., & Pessôa, V. L. S. (2006). *Assentamentos rurais e as transformações territoriais no Pampa Gaúcho/RS/BR*. Conference proceedings of 6ª Biennial Colloquium of Territorial Transformations. UNL.
Cotton, M., & Devine-Wright, P. (2012). Making electricity networks "visible": Industry actor representation of "publics" and public engagement in infrastructure planning. *Public Understanding of Science, 21*(1), 17–35.
Cowell, R., Bristow, G., & Munday, M. (2011). Acceptance, acceptability and environmental justice: The role of community benefits in wind energy development. *Journal of Environmental Planning and Management, 54*(4), 539–557.
de Sousa Santos, B. (2010a). *Descolonizar el saber, reinventar el poder*. Trilce-Extensión Universitaria.
de Sousa Santos, B. (2010b). *Una epistemología del Sur*. Siglo XXI.
Devine-Wright, P. (2005). Local Aspects of UK renewable energy development: Exploring public beliefs and policy implications. *Local Economy, 10*, 57–69.
Devine-Wright, P. (2009). Rethinking NIMBYism: The role of place attachment and place identity in explaining place-protective action. *Journal of Community & Applied Social Psychology, 19*, 426–441.
Devine-Wright, P., & Howes, Y. (2010). Disruption to place attachment and the protection of restorative environments: A wind energy case study. *Journal of Environmental Psychology, 30*(3), 271–280.
Di Masso, A., Dixon, J., & Durrheim, K. (2014). Place attachment as discursive practice. In L. C. Manzo & P. Devine-Wright (Eds.), *Place attachment: Advances in theory, methods and applications* (pp. 75–86). Routledge.
Di Masso, A., Dixon, J., & Hernandez, B. (2017). Place attachment, sense of belonging and the micro-politics of place satisfaction. In G. Fleury-Bahi, E. Pol, & O. Navarro (Eds.), *Handbook of environmental psychology and quality of live research* (pp. 85–106). Springer.
Dixon, J., & Durrheim, K. (2000). Displacing place-identity: A discursive approach to locating self and other. *British Journal of Social Psychology, 39*, 27–44.

Farias, T. M. (2017). *Afetividade e resistência: vínculo, transformações socioambientais e oposição capital-lugar na cidade de Galinhos-RN*. (Affectivity and resistance: Bond, socioenvironmental transformations and capital-place opposition in the city of Galinhos-RN). Doctoral dissertation. https://repositorio.ufrn.br/jspui/bitstream/123456789/24345/1/TadeuMattosFarias_TESE.pdf

Ferreira, J. P. (1959). *Enciclopédia dos municípios brasileiros* (vol. XXXIII). http://biblioteca.ibge.gov.br/visualizacao/livros/liv27295_34.pdf

Finley-Brook, M., Williams, T. L., Caron-Sheppard, J. A., & Jaromin; M. K. (2008). Critical energy justice in US natural gas infrastructuring. *Energy Research & Social Science, 41*, 176–190.

Finley-Brook, M., & Thomas, C. (2011). Renewable energy and human rights violations: Illustrative cases from indigenous territories in Panama. *Annals of the Association of American Geographers, 101*(4), 863–872.

Groves, C. (2015). The bomb in my backyard, the serpent in my house: Environmental justice, risk, and the colonization of attachment. *Environmental Politics, 24*(6), 853–873.

Haesbaert, R. (1988). *RS: Latifúndio e Identidade Regional.* Mercado Aberto.

Howe, C. (2014). Anthropocenic ecoauthority: The winds of Oaxaca. *Anthropological Quarterly, 87*(2), 381–401.

IBGE Instituto Brasileiro de Geografia e Estatística. (2010). *Santana do Livramento—Censo Demográfico.* http://www.cidades.ibge.gov.br/v4/brasil/rs/santana-do-livramento/panorama

Kaiser, J. (1999). *Ordem e progresso: O Brasil dos gaúchos.* Insular.

Lenoir-Improta, R. (2017). *Turbines in the Pampa: The role of place meanings and people-place bonds in the acceptance and legitimation of a new energy infrastructure in Brazil.* Unpublished doctoral thesis. University of Barcelona, Barcelona, Spain.

Lenoir-Improta, R., & Pinheiro, J. Q. (2011). Socio-environmental impacts of Brazil's first large-scale wind farm. In P. Devine-Wright (Ed.), *Renewable energy and the public: From NIMBY to participation* (pp. 219–231). Earthscan.

Mouffe, C. (2013). *Agonistics: Thinking the world politically.* Verso Books.

Pol, E., Di Masso, A., Castrechini, A., Bonet, M. R., & Vidal, T. (2006). Psychological parameters to understand and manage the NIMBY effect. *Revue Européenne de Psychologie Appliquée, 56*, 43–51.

Quijano, A. (2000). Colonialidad del poder, eurocentrismo y América Latina. In E. Lander (Ed.), *La colonialidad del saber: eurocentrismo y ciencias sociales. Perspectivas latinoamericanas.* CLACSO.

Scheer, H. (2009). *Autonomía Energética: la situación económica, social y tecnológica de la energía renovable.* Icaria.

Späth, P. (2012). Understanding the social dynamics of energy regions—the importance of discourse analysis. *Sustainability, 4*, 1256–1273.

Swyngedouw, E. (2010). Apocalypse forever? Post-political populism and the spectre of climate change. *Theory, Culture & Society, 27*(2–3), 213–232.

Walker, G. (2009). Beyond distribution and proximity: Exploring the multiple spatialities of environmental justice. *Antipode, 41*(4), 614–636.

Walker, G. (2012). *Environmental justice: Concepts, evidence and politics*. Routledge.

Wüstenhagen, R., Wolsink, M., & Bürer, M. J. (2007). Social acceptance of renewable energy innovation: An introduction to the concept. *Energy Policy, 35*, 2683–2691.

Zografos, C., & Martínez-Alier, J. (2009). The politics landscape value: A case study of wind farm conflict in Catalonia. *Environmental and Planning A, 41*, 1726–1744.

CHAPTER 12

Energy Justice and Social Acceptance of Renewable Energy Projects in the Global South

Dan van der Horst, Rebecca Grant, Adolfo Mejía Montero, and Aiste Garneviciene

Introduction

We posit that social acceptance can largely (but not exclusively) be unpacked as a range of questions about fairness; e.g. about the distribution of benefits and negative impacts, the appropriate use of tax-payer money, the protection of local public goods threatened by development, the need to mitigate the threats of climate change, and the processes of decision making (Wüstenhagen et al., 2007; Batel et al., 2013; Upham et al., 2015). By logical extension, this would suggest that recent scholarship on Energy Justice (EJ) could be used to analyse public responses to energy projects. The aim of this chapter is to ask, by examining case

D. van der Horst (✉) • R. Grant • A. M. Montero
The University of Edinburgh, Edinburgh, UK
e-mail: Dan.vanderHorst@ed.ac.uk

A. Garneviciene
Ofgem, London, UK

S. Batel, D. Rudolph (eds.), *A critical approach to the social acceptance of renewable energy infrastructures*,
https://doi.org/10.1007/978-3-030-73699-6_12

studies in the Global South (GS), (how) does EJ help us to understand observed differences in the perceptions of renewable energy (RE) projects? The case studies stem from MSc and PhD dissertation fieldwork carried out individually by three of the co-authors and reported in more detail elsewhere (Garneviciene, 2019; Grant, 2019; Mejia-Montero, 2020). For this chapter we collaborated to reanalyse these case studies through an EJ lens, individually and collectively. The structure and content of the chapter grew 'organically' out of an iterative set of discussions between the four co-authors during the Covid-19 lockdown period. We note as caveats that both the social acceptance and EJ literatures originated in the Global North (GN) to study issues in the GN, though our empirical basis draws on three case studies in the GS. We note too that the use of 'GS' to group our case studies is imperfect, risking the homogenisation of countries and context by geographic binary.

The chapter is structured as follows. First, we provide a short overview of the origins and analytical focus of energy justice. We then apply the EJ framework (EJF) to three case studies differing in scale (grid/off-grid), technology (wind/solar PV/hydropower) and ownership (commercial/social enterprise) in Mexico and Malawi. Finally, in our discussion we reflect on what the EJF can reveal about differences in social acceptance between these individual case studies, and between the GS and GN.

Energy Justice; What It Is and Where It Came From

EJ scholarship has emerged in the last decade in response to increasing focus from social scientists on the impacts of the energy transition (Heffron et al., 2015; Jenkins et al., 2016; Sovacool & Dworkin, 2015). In origin, EJ could be characterised as a thematic academic specialisation of environmental justice scholarship. However, environmental justice itself first emerged as a local, grassroots and activist led movement at the nexus of the civil rights movement and the environmental movement in the US (Bullard & Johnson, 2000; Holifield et al., 2009). In contrast, Jenkins (2018) argues that EJ is distant from these anti-establishment origins which, coupled with its narrower thematic focus (on energy systems), could equip EJ to be more salient to those in policy (see also Sovacool et al., 2017).

Different analytical frameworks have emerged to better understand EJ since its inception, including the Energy Justice Framework (EJF) which conceptualises justice into distributional, procedural and recognition tenets (McCauley et al., 2013, Jenkins et al., 2016). This framework is both evaluative and normative, seeking to address the following questions: '(a) where injustices emerge, (b) which affected sections of society are ignored, (c) which processes exist for their remediation in order to (i) reveal, and (ii) reduce such injustices' (Jenkins et al., 2016, p. 175). More specifically, distributional justice refers to the distribution of benefits and risks (who benefits/loses and by how much) across energy supply chains. In turn, the recognition justice tenet seeks to identify who is ignored (non-recognition) and misrepresented (misrecognition). Procedural justice refers to transparency, legitimacy and fairness in decision making processes, necessitating inclusive and appropriate stakeholder involvement (ibid.; Heffron & McCauley, 2014). These tenets are derived from earlier environmental justice writings and they draw on the seminal works of philosophers of justice including Rawls and Fraser (Heffron et al., 2015; Jenkins et al., 2016).

The empirical contexts for EJ research are starting to expand beyond the GN. In settings where a functional national grid cannot be taken for granted, recent EJ related research has focused on the distributive questions of who gets connected, how this is decided, and the mixed experiences of first-time electricity users (Boamah & Rothfuß, 2020; Tarekegne, 2020; Monyei, Adewumi et al., 2018; Monyei, Jenkins et al., 2018). Drawing on this momentum, we brought together three case studies representing three different grid situations: a mature national grid providing almost full national coverage with the aid of wind power in Mexico, a black-out prone and incomplete national grid boosted by a new solar farm which bypasses local communities in Malawi, and a new rural micro-hydro mini-grid providing low level electrification in Malawi.

Case Study 1: On-grid Wind Power in Oaxaca, Mexico

We first examine energy (in)justices related to the promotion and construction of many wind farms on the Isthmus of Tehuantepec, Oaxaca, Mexico. Our findings draw on 58 in-depth interviews with members of opposition groups, landowners, project developers and public servants, conducted by Adolfo Mejia-Montero (co-author) between October 2017 and September 2019.

The Mexican revolution of 1910 saw the widespread redistribution of land to agrarian and indigenous communities (61% of the Isthmus population), and in Oaxaca 77.6% of land was put under collective ownership (Barbary, 2015; Gobierno de Oaxaca, 2011). However, since the 1970s, collective local institutions in the Isthmus have been eroded, generating conflict surrounding land ownership (Oceransky, 2008). The promotion of wind power in Mexico appeals to both discourses of national decarbonisation and regional socio-economic development for this relatively disadvantaged region (Borja-Diaz et al., 2005). Nevertheless, authors including Zárate Toledo and Fraga (2019), Dunlap (2019), and Velasco-Herrejon and Savaresi (2019), are critical of this promotion, highlighting the unequal distribution of socio-economic benefits and concerns over local environmental degradation. Furthermore, the arrival of (predominantly Spanish) transnational wind power companies, coupled with historical colonisation in the Isthmus region, has provided local resistance groups with new narratives of conquest where wind power is a tool of neoliberalism (Boyer, 2019).

In Oaxaca, distributive injustices are evidenced in the unequal distribution of socio-economic benefits coming from RET. For example, most electricity generated by wind farms is consumed by large private companies outside the region despite some local populations lacking high quality energy services (Zárate Toledo & Fraga, 2019). This was highlighted by interviewees as problematic given existing concerns over safety (especially for women) during power outages and the status of wind as collective resource; "(it is) not fair that on our farms many times we have no power, or we have power cuts, when we produce loads of power on our own lands!" (farmer, Union Hidalgo). Furthermore, despite promises of regional socio-economic development, manufacturing of turbine components took place almost exclusively in the United States and in Europe with limited local job creation (Huesca-Pérez et al., 2016).

Interviewees described a set of procedural injustices in the early stages of wind power development in the region, frequently linked to the lack of formal consultation processes following the principles of Free, Prior and Informed Consent (Nahmad-Sittón et al., 2014). Colloquiums, set up by the Mexican Government to discuss the development of guidelines relating to wind power, initially lacked local and indigenous voices, whilst government and wind power companies were over-represented (Oaxaca Economy Ministry, 2010; Cruz-Velázquez, 2008; Boyer, 2019). Meetings prioritised developing a legal system to promote the techno-economic feasibility of wind power, omitting potential adverse impacts defined by excluded stakeholders. Furthermore, procedural justice applies also to the responsibility to deliver benefits, and accountability for failure to do so in settings where multiple actors are implicated in development. Interviewees indicated there was confusion over whether it was the role of the state or that of developers to build a manufacturing industry; "People claim [...] they were told there would be work. This puts us (project developers) in a difficult position. But that must be a state policy, the government should be the promoter our company only sells electricity" (developer, Juchitan).

Closely linked to the lack of fair and meaningful inclusion of all stakeholders in decision making identified above, recognition injustices also related to (1) non-recognition of non-humans and (2) non-recognition and misrecognition of local and indigenous populations. One opposition leader highlighted concern for the neglect of the impacts of wind power on ecosystem functionality; "We welcome technology, we welcome wind power, but the one which does not harm groundwater or the flora and fauna of my region, and that gives something in return to local communities". This tension reflects more widely on the burden carried by indigenous communities (who themselves have not contributed to climate change) to accept the local siting of RE power plants. Moreover, community members who lacked land in sites deemed strategic for wind power articulated concerns over a lack of recognition (in both formal consultation and distribution of benefits) in contrast to others; "They (wind power companies) have violated our indigenous rights...they should be the ones to compensate us and pay ... There are a lot of benefits for companies, and some for landowners and municipal governments, but for society? There are no benefits at all!" (environmental activist, Union Hidalgo).

Recognising that concerns identified above have been highlighted in other work, we argue that reframing through the EJF holds value as a way to understand social acceptance as negotiated, contested and nuanced. Specifically, the EJF highlights a need to look at recognition justice (and what this means in practice) for both human and non-human stakeholders as a condition of acceptance (Sovacool et al., 2017).

Case Studies 2a and 2b: Hydropower and Solar PV Systems in Malawi

Electricity access in Malawi remains low, with only 18% of the country's population having access to electricity in 2018 (World Bank, 2020). Furthermore, insufficient generation, transmission and distribution facilities means that even those connected to the grid experience frequent and often prolonged power outages. A reliance on the Shire river for electricity increases vulnerability, and in periods of drought up to 90% (362.8 MW) of Malawi's total electricity supply is lost. However, the Government of Malawi (GoM) identifies access to electricity as essential for economic growth and expanding access is a priority under the new National Energy Policy. Inviting private sector participation in electricity generation, this framework aims to electrify 80% of the country by 2035 (Kachaje et al., 2017; GoM, 2018).

2a: Hydro Mini-grid in Mulanje

The hydro mini-grid in Mulanje is a social enterprise, run by the Mulanje Electricity Generation Agency (MEGA) since 2014 and funded through the Mount Mulanje Conservation Trust (majority owner), the charity Practical Action and the Scottish Government (Wood Group UK, 2018; Eales, 2018). It aims to provide low-cost electrification to a region currently unconnected to the national grid. Financial sustainability is pursued through a tiered tariff, with higher business rates subsidising lower domestic and social consumer rates (Kadziponye, 2019).

18 semi-structured interviews were carried out by Rebecca Grant (co-author) in March and June 2019. These included 10 with representatives from non-governmental organisations and electricity regulatory authorities in Malawi (#1–10), and 8 with local energy users of the off-grid hydro mini-grid in Mulanje (#11–18), conducted with the help of a translator provided through MEGA. For more details, see Grant (2019).

Together, recognition and procedural justice call for the recognition and inclusion of all affected stakeholder groups in appropriate discussions prior to, during, and post-installation of energy systems. Recognition must extend to those presently without electricity, whose aspirations and concerns should be considered discussions of electrification. Dominant economic narratives regarding cost effectiveness and least cost electrification inevitably shape who has access to electricity within Malawi. Some expert interviewees dismissed the need for rural groups across Malawi, who traditionally did not have electricity, to be incorporated into the sector: "these people have lived for hundreds of years without electricity-why bother now" (#7). The biggest barrier to electricity access in rural areas was 'ability to pay' (#3), which could be aggregated into a market potential: "(they are a) small market…not big enough to be of interest to developers" (#10). These narratives can be reductive; understanding energy use only through assumptions by dominant voices about local people's current ability to pay is problematic as an implicit measure for 'deserving' access. This is a form of misrecognition as it disregards the individual and collective voices, aspirations and economic potential of unelectrified local populations. It also presents a static vision of the world where both costs and fortunes are assumed to be fixed into the future, and where social and economic models of delayed return on (economic or social) investment are not considered.

The presence of a waiting list for connection to the MEGA managed mini-grid, in addition to conversations with local energy users, indicates that the benefits of electrification are extensive even within communities that are considered low income (and by default low demand). Echoing Matinga and Annegarn (2013) in their recognition of electricity as enabling global connection through electronic use, energy users noted benefits through being able to watch TV and listen to the radio (#12, 15, 16). Connection to the mini-grid was described as a pull for talented labour such as doctors and teachers (#13) to schools and hospitals; "electricity…attracts outside attention" (#12). The social acceptance of mini-grids is thus tied to their capacity to connect, both within communities (e.g. connecting consumers to services) and between communities (e.g. in attracting skilled labour and in enabling connection to local and global networks). A lack of access to electricity for some (e.g. those on waiting lists) and the inability of electricity to meet connection needs of all (through constraints to generation) are also metrics through which social acceptance is defined post-installation of off-grid technologies.

Legislative and regulatory systems were described as hindering the roll-out of a least cost electrification plan in parts of rural Malawi, impacting those without access to electricity and entrenching a system of "seemingly randomly allocated mini-grids" (#10). Interviewee #9 summarised foreign investor perspectives as follows; "on—grid will focus on big towns…' and only 'once the grid is mature… off-grid developers will be more comfortable". An investment environment described as 'risky', coupled with the only recent decoupling of ESCOM and EGENCO which opened the market to independent power producers (IPPs), detracts for-profit off-grid investors in Malawi. The siting of the mini-grid system only advantaged those whose "house is within the grid network where it is easier for the drop wire to reach the house" (#15). There is an injustice of geographical circumstance where those who live just outside of the catchment for grid electricity are 'arbitrarily disadvantaged in their ability to attain essential energy services' (Bouzarovski & Simcock, 2017, p. 645).

Concerns were also raised over the siltation of the micro-hydro grid through deforestation for fuel, wood and charcoal on the surrounding slopes; "with deforestation there is siltation… generation gets lower and lower" (#8). The impacts of deforestation upstream limited provision of electricity to downstream users. Whilst interviews indicated the decreasing use of biomass for lighting, demand for fuel wood remained high for cooking, even in households connected to the mini-grid. The promotion of RE and transition to electrification (for lighting) does not offset reliance on biomass (Taulo et al., 2015; Hiemstra-van der Horst & Hovorka, 2008). Understanding how these two systems interact, and the extent to which the promotion of RE systems with low generation capacity addresses the spectrum of energy needs, is important to fully understand local social acceptance and to allow a more meaningful comparison with the perspectives of on-grid local residents in other case studies.

2b: Grid-Connected PV Plant in Salima

The second case study focuses on Salima Solar, a 60MW solar PV plant owned and managed by the ProjectCo consortium and the first Independent Power Producer (IPP) scheme in Malawi to reach construction phase (ProjectCo, 2019). Fieldwork was undertaken by Aiste Garneviciene (co-author), with facilitation and translation by Dr Nancy Chawawa, in May 2019 to examine risks and benefits of this development. As part of this, 25 interviews were carried out (15 with affected local community members and 9 with district officials, ProjectCo and the Acting Director for the Department of Energy Affairs) in addition to focus group meetings with chiefs and local community members as part of the fieldwork (see Garneviciene, 2019).

Once operational, Salima solar plant is expected to almost double Malawi's electricity generation and reduce load shedding on the national grid. However, the promised energy security bypasses local communities who are not grid connected and will not receive electricity access through this project. Individual households instead 'benefit' through monetary compensation for land acquisition. Acceptance of the project was high among those who received higher payments for the land; in some cases over 1 million MWK (£1050) per acre of land (ProjectCo, 2019). As the project is in an early stage, for a community with no previous experience with other RET projects, Salima solar project represents hope for economic development in the region, as well as opportunities to (re)invest the compensation.

High initial acceptance of the project is also linked to high levels of engagement with local communities in consultation and decision making. Project developers visited the communities regularly and, in villages that operate according to strict hierarchies (McNamara, 2019), legitimacy was strengthened by the involvement of district council officials and support from the Traditional Authority (TA) Senior Chief Kalonga who facilitated land acquisition. However, whilst Chief Kalonga noted that "all the villages were very happy because the project was implemented", concerns were raised over a lack of binding agreements validating ProjectCo promises to build a school and a hospital in Kazimbe. Kalonga also expressed offence that procedures required by tradition were not followed and he has not received a "Zikomo" (thank you) for his involvement. There is evidence of both procedural justice (in engagement according to local protocols) and procedural injustices through failure to ensure accountability and lack of *meaningful* and *appropriate* inclusion of stakeholders within the same setting. Procedural justice is thus not static or one-dimensional, a lesson transferable to assessments of acceptance.

An Environmental and Social Impact Assessment (ESIA) anticipates that local communities (subsistence farmers with few households generating additional income) will be severely impacted (ProjectCo, 2019). Land loss from land acquisition is also expected to aggravate food insecurity and increase poverty. Compensatory payments to individuals for land acquisition which differed according to size of plot owned and the time land was acquired, were also deemed unequal and unfair by some. This has also been observed in other top-down RET projects in the GS, in India (Yenneti & Day, 2016) and understood as a distributive injustice in Oaxaca (Mejia-Montero, 2020). Concerns were also raised over the potential use of compensation as a mechanism to deter criticism of the project locally (given high payments), although ProjectCo assured that the process was fair;. "compensation approach was not to silence them, but to make their lives…better" Due attention should be paid to the risk of monetary compensation being conflated with just process as a source of procedural injustice. Unequal distribution (perceived or real) of benefits from compensation coupled with the burden of food insecurity for current and future generations, also constitute distributive injustices in this case study.

Furthermore, social acceptance is conditional and dynamic; expectations of local electricity access emerged following initial improvements to livelihoods. This assumption was shared not only by individual households but also by Chief Kolonga who noted that "electricity was promised" and "communities will rise" if this was not met. This dissonance between promised outcomes (with ProjectCo policy clear on contribution to national grid) and expectations reflects potential procedural injustices relating to insufficient or inappropriate articulation of benefits and risks, and a lack of due consideration and communication of short and long term impacts.

Salima solar was promoted and framed as being of 'national interest' by the former District Environmental Development Officer, enhancing the energy security of the already grid-connected part of Malawi, but not for those living off-grid. Separately, the district commissioner also noted that the community knew that electricity generated by the Salima solar plant would be exported elsewhere but "understood the problem of the nation". Appeals to national interest in assessing local acceptance (where the project is renewable and feeds into the national grid) by some 'elite' stakeholders risks silencing local stakeholders whose concerns or critiques relate to local impacts.

Discussion

This chapter has examined the key Energy Justice implications of three renewable energy projects in the Global South. Our tentative findings explore the similarities and differences between the three case studies and seek to link these to the existing social acceptance literature in the Global North. However, before we discuss these, it is important to recognise that the authors, with their case study materials, have come together post-hoc to develop this chapter. Differences in methodology and data collection efforts across the three case studies pose limits to comparative analysis. As with all research conducted outside of the researcher's lived experiences, positionality and undeserved privileges influence what is articulated, to whom and how (see Sultana, 2007). We cannot claim to have captured the full diversity and depth of different voices. However, we would argue that our choice of case studies; full grid, partial grid and mini-grid, are sufficiently 'archetypical' to be able to draw some tentative conclusions.

Commonalities; the Role of Temporality in Shaping Acceptance

Comparative analysis drew our attention to the importance of the temporal stages of projects in shaping acceptance. Social acceptance seemed to be much higher in Salima than in Oaxaca (both grid projects without local electrification benefits), but the former was in the construction phase with fewer opportunities for negative impacts to emerge. In Oaxaca, wind farms have been operational since 2007, with disappointments emerging due to limited job opportunities and due to observed environmental impacts. In Salima, positive perceptions may be linked to unrealistic expectations, especially of grid access (evidenced already in assumptions of connection). Therein lies the risk of future disappointment, as realised in the grid connected Lake Turkana Wind Farm where acceptance in local communities seemed high initially, though later research revealed significant concerns of unjust processes and uneven distribution of benefits (Cormack & Kurewa, 2018). However, in Mulanje, social acceptance is still high five years after the project became operational. We think this is due to fulfilled aspirations of access to electricity, and ongoing local engagement; MEGA is locally based, employs a largely local workforce and is continuing to invest in expansion of local generation capacity in response to local demand articulated through regular meetings held with a committee of representatives from the area affected (MEGA, 2019). This local social enterprise model also benefits from comparison with neighbouring rural communities where access to electricity remains non-existent.

Differences between local perceptions during early engagement and post-process stages have been noted in the social acceptance literature (Wolsink, 2007), but also highlighted in recent EJ literature, describing the importance of 'expectation management' (Dwyer & Bidwell, 2019). Case studies point to a need to consider, at the inception of developments, how the EJ tenets can best be applied to capture and stay attuned to conditions of social acceptance, whilst these are diverse and non-static throughout the lifecycle of RETs.

Divergences; Understanding Social Acceptance with Respect to Existing Electrification

Our case studies illustrate the substantial difference between communities that are grid connected and those that are not with respect to social acceptance. In Oaxaca, where grid connection is ubiquitous, many of the concerns raised by local residents sound familiar to the ones reported in the GN, e.g. impacts of construction, the lack of local manufacturing and jobs, the capture of financial benefits by big landowners. Parallels can be drawn between concerns over exploitation of common, biodiverse land in Oaxaca and protester narratives in the GN regarding the protection of 'unspoiled' common land (e.g. Woods, 2003). Narratives portraying foreign companies through a lens of historical injustice (Spanish companies as (re)conquistadores) also share conceptual resonance with the dismissal of commercial project developers as a form of conquest by a historic foreign foe in the GN (e.g. Ellis et al., 2007).

Routes for formal expression and articulation of concerns and needs relating to RE also differ across case studies, necessitating flexible conceptualisations of what procedural justice looks like in pre- and post-electrification settings. In Oaxaca, an initial failure of the Mexican government and wind power developers to recognise indigenous rights led to sustained local resistance and the development of Free, Prior and Informed Consent mechanism and Social Impact Assessments (though with local concerns over their use being conflated with acceptance) (Mejia-Montero et al., 2020; Dunlap, 2017). In off-grid locations, people do not 'disconnect' their acceptance of supply side technologies from their aspirations for energy services. In Malawi then, it is also essential to ensure that self-defined energy use aspirations are articulated to protect the rights of all to safe, sufficient, affordable and accessible energy (the 'rights to the grid'), where access is not the default.

Disentangling Procedural and Recognition (In)Justice in Practice

Misrecognition and non-recognition of different stakeholders was evidenced in all three case studies, though articulated differently. Misrecognition manifests across different scales, as evidenced in Malawian case studies. In Mulanje, 'elite' stakeholders working at national levels reduced future energy users to their market potential (in being low income and low demand) in discourses of deservedness of electrification. Promoting RE through appeals to national interest in Salima risks homogenising local end users as one 'community' rather than diverse stakeholders with differentiated interests. Observed fluctuations in payment levels over time (for land bought or leased, or as community benefit-sharing) reveal procedural injustices (in unequal engagement and lack of transparency when negotiating with stakeholders), distributional injustices (whereby some stakeholders benefit more than others in the development of solar PV) and recognition injustices (with respect to whose concerns are articulated and how). As a critique of the validity of EJ tenets, the boundaries between the recognition and procedural justice can be particularly blurred in the GS, due to tensions between formal state policies (which may look good on paper) and what happens on the ground when powerful interests interact with indigenous communities whose views and expectations are shaped by traditional belief systems, informal institutions and collective memories of colonial exploitation.

Distributive Justice and Defining 'Elite' Capture Across Different Settings

The final cross-cutting theme is the apparent elite capture of benefits. This chimes with recent work on EJ discussing how low-carbon transitions create, perpetuate, challenge or entrench the power of elites (Sovacool & Brisbois, 2019). In Oaxaca, the possession of private land, privileged information and political influence by elite stakeholders were weaponised to secure unfair access to benefits from the wind project (Alonso-Serna & Mejia-Montero, 2019; Juárez-Hernández & León, 2014; Nahmad-Sittón et al., 2014). In Salima, the export of electricity access to the national grid, alongside the unequal distribution of benefits through compensation, could be conceptualised as 'elite' capture given the relatively small groups that benefit and the high level of locally unmet expectations of connection. In Mulanje, interviews indicated that those with homes deemed safe for wiring (ability to wire own homes and to have these certified by an engineer), were able to gain access to electricity prior to others. With current limitations to generation, and the presence of a waiting list, the 'elite' capture of benefit may refer to ability to utilise electricity as a result of previous home ownership, wealth and status. There is a powerful visible symbolism tied to connection to a physical grid, as well as having a home suitable for connection, as indicative of not only wealth but also investment into communities and households (see Boamah, 2020; Boamah & Rothfuß, 2020). Recognising physical grid connection as both provisioning and symbolic is essential in assessing the acceptance of RE technologies in previously unelectrified settings (and in how we might recognise emerging distributive injustices).

Together this chapter assesses the utility of EJ to understand social acceptance across three case studies in the GS. The three tenets of EJ help to highlight the complex ways in which energy (in)justices emerge in both grid-connected and unconnected settings, demanding social acceptance to be contextually rooted. It also points to the need to consider whose opinions and insights shape discourses of acceptance in the GS, how we best recognise and reflect on the needs of populations who have experienced marginalisation, and the need for temporal reflexivity with social acceptance a changing state. Whilst some of the key findings from social acceptance and EJ literature in the GN may appear to resonate in the GS (and vice versa, now that the literature from the GS is growing), it is important to uncover and understand the many nuances and conditionalities. But we also find stark differences and special attention must be afforded to the diverse and dynamic experiences and expectations of those in pre-electrified and newly electrified settings.

References

Alonso-Serna, L., & Mejia-Montero, A. (2019). Rentas eólicas y nuevos procesos de diferenciación social en el Istmo de Tehuantepec, Oaxaca. In C. Tornel (Ed.), *Alternativas para limitar el calentamiento global en 1.5°C Más allá de la economía verde* (1st ed., pp. 269–292). Fundación Heinrich Böll

Barbary, O. (2015). Social inequalities and indigenous populations in Mexico: A plural approach. In *Social statistics and ethnic diversity—cross-national perspectives in classifications and identity politics* (pp. 209–228). Springer Open.

Batel, S., Devine-Wright, P., & Tangeland, T. (2013). Social acceptance of low carbon energy and associated infrastructures: A critical discussion. *Energy Policy, 58*, 1–5. https://doi.org/10.1016/j.enpol.2013.03.018

Boamah, F. (2020). Desirable or debatable? Putting Africa's decentralised solar energy futures in context. *Energy Research & Social Science, 62*, 101390. https://doi.org/10.1016/j.erss.2019.101390

Boamah, F., & Rothfuß, E. (2020). 'Practical recognition' as a suitable pathway for researching just energy futures: Seeing like a 'modern' electricity user in Ghana. *Energy Research & Social Science, 60*, 101324. https://doi.org/10.1016/j.erss.2019.101324

Borja-Diaz, M. A., Jaramillo-Salgado, O. A., & Mimiaga-Sosa, F. (2005). Primer Documento del Proyecto Eoloélectrico del corredor Eólico del Istmo de Tehuantepec, Elementos de una Estrategia de Promoción de la generación de Electricidad a partir de energía renovable. https://www.ier.unam.mx/~ojs/pub/Eolica/LibroProyectoEolico/ProyectoEolicoOaxaca_v1.pdf

Bouzarovski, S., & Simcock, N. (2017). Spatializing energy justice. *Energy Policy, 107*, 640–648. https://doi.org/10.1016/j.enpol.2017.03.064

Boyer, D. (2019). *Energopolitics: Wind and power in the Anthropocene*. Duke University Press.

Bullard, R. D., & Johnson, G. S. (2000). Environmentalism and public policy: Environmental justice: Grassroots activism and its impact on public policy decision making. *Journal of Social Issues, 56*(3), 555–578. https://doi.org/10.1111/0022-4537.00184

Cormack, Z., & Kurewa, A. (2018). The changing value of land in Northern Kenya: the case of Lake Turkana Wind Power. *Critical African Studies, 10*(1), 89–107. https://doi.org/10.1080/21681392.2018.1470017

Cruz-Velázquez, B. (2008). Desarrollo regional en el Istmo de Tehuantepec: Una Perspectiva Desde el Territorio. *Aquí estamos, 5*(8), 7–16.

Dunlap, A. (2017). "A bureaucratic trap:" Free, Prior and Informed Consent (FPIC) and wind energy development in Juchitán, Mexico. *Capitalism Nature Socialism, 29*(4), 88–108. https://doi.org/10.1080/10455752.2017.1334219

Dunlap, A. (2019). *Renewing destruction: Wind energy development, conflict and resistance in a Latin American context* (p. 205). Rowman & Littlefield International.

Dwyer, J., & Bidwell, D. (2019). Chains of trust: Energy justice, public engagement, and the first offshore wind farm in the United States. *Energy Research & Social Science, 47*, 166–176. https://doi.org/10.1016/j.erss.2018.08.019

Eales, A. (2018). *Mini-grids in Malawi: Status, opportunities and barriers*. https://www.researchgate.net/publication/326926237_Mini-grids_in_Malawi_Status_Opportunities_and_Barriers

Ellis, G., Barry, J., & Robinson, C. (2007). Many ways to say 'no', different ways to say 'yes': Applying Q-methodology to understand public acceptance of wind farm proposals. *Journal of Environmental Planning and Management, 50*(4), 517–551. https://doi.org/10.1080/09640560701402075

Garneviciene, A. (2019). *Reimagining Salima solar: sociotechnical imaginaries of the first utility-scale solar project in Malawi*. MSc thesis, University of Edinburgh.

Gobierno de Oxaca (2011). Planes Regionales de Desarrollo de Oaxaca 2011-2016 - Istmo. Oaxaca. Available at: https://www.finanzasoaxaca.gob.mx/pdf/planes/planes_regionales/2011-2016/Istmo.pdf.

Government of Malawi. (2018). Malawi National Energy Policy. Ministry of Natural Resources, Energy and Mining. https://www.energy.gov.mw/index.php/blog/downloads/policies-strategies

Grant, R. (2019). *A combined energy justice and ecohealth approach to understanding rural energy use in Malawi*. MRes thesis, University of St Andrews.

Heffron, R. J., & McCauley, D. (2014). Achieving sustainable supply chains through energy justice. *Applied Energy, 123*, 435–437. https://doi.org/10.1016/j.apenergy.2013.12.034

Heffron, R. J., McCauley, D., & Sovacool, B. K. (2015). Resolving society's energy trilemma through the energy justice metric. *Energy Policy, 87*, 168–176. https://doi.org/10.1016/j.enpol.2015.08.033

Hiemstra-van der Horst, G., & Hovorka, A. J. (2008). Reassessing the "energy ladder": Household energy use in Maun, Botswana. *Energy Policy, 36*(9), 3333–3344. https://doi.org/10.1016/j.enpol.2008.05.006

Holifield, R., Porter, M., & Walker, G. (2009). Introduction spaces of environmental justice: Frameworks for critical engagement. *Antipode, 41*(4), 591–612. https://doi.org/10.1111/j.1467-8330.2009.00690.x

Huesca-Pérez, M. E., Sheinbaum-Pardo, C., & Köppel, J. (2016). Social implications of siting wind energy in a disadvantaged region—The case of the Isthmus of Tehuantepec, Mexico. *Renewable and Sustainable Energy Reviews, 58*, 952–965. https://doi.org/10.1016/j.rser.2015.12.310

Jenkins, K. (2018). Setting energy justice apart from the crowd: Lessons from environmental and climate justice. *Energy Research & Social Science, 39*, 117–121. https://doi.org/10.1016/j.erss.2017.11.015

Jenkins, K., McCauley, D., Heffron, R., Stephan, H., & Rehner, R. (2016). Energy justice: A conceptual review. *Energy Research & Social Science, 11*, 174–182. https://doi.org/10.1016/j.erss.2015.10.004

Juárez-Hernández, S., & León, G. (2014). Energía eólica en el istmo de Tehuantepec: desarrollo, actores y oposición social. *Problemas Del Desarrollo, 45*(178), 139–162. https://doi.org/10.1016/s0301-7036(14)70879-x

Kachaje, O., Chisanu, L., & Liangjun, Y. (2017). Electricity reforms in Malawi; impacts and consequences. *International Journal of Renewable Energy Technology Research, 6*(4), 1–9.

Kadziponye, A. (2019). *MMCT/MUREA/MEGA Malawi energy access programme [powerpoint presentation]*. MEGA.

Matinga, M. N., & Annegarn, H. J. (2013). Paradoxical impacts of electricity on life in a rural South African village. *Energy Policy, 58*, 295–302. https://doi.org/10.1016/j.enpol.2013.03.016

McCauley, D. A., Heffron, R. J., Stephan, H., & Jenkins, K. (2013). Advancing energy justice: the triumvirate of tenets. *International Energy Law Review, 32*(3), 107–110. http://hdl.handle.net/1893/1834

McNamara, T. (2019). The limits of Malawian headmen's agency in co-constructed development practice and narratives. *Journal of Eastern African Studies, 13*(3), 465–484. https://doi.org/10.1080/17531055.2019.1599196

MEGA. (2019). *About us.* MEGA. https://www.mega.mw/about-us

Mejia-Montero, A. (2020). *Energy justice and utility-scale wind power in the isthmus: Context, resistance and identity.* PhD thesis, University of Edinburgh.

Mejia-Montero, A., Alonso-Serna, L., & Altamirano-Allende, C. (2020). The role of social resistance in shaping energy transition policy in Mexico: The case of wind power in Oaxaca. In L. N. Guimaraes (Ed.), *The regulation and policy of Latin American energy transitions* (pp. 397–415). Elsevier.

Monyei, C. G., Adewumi, A. O., & Jenkins, K. E. H. (2018). Energy (in)justice in off-grid rural electrification policy: South Africa in focus. *Energy Research & Social Science, 44*, 152–171. https://doi.org/10.1016/j.erss.2018.05.002

Monyei, C. G., Jenkins, K., Serestina, V., & Adewumi, A. O. (2018). Examining energy sufficiency and energy mobility in the global south through the energy justice framework. *Energy Policy, 119*, 68–76. https://doi.org/10.1016/j.enpol.2018.04.02

Nahmad-Sittón, S., Nahón, A., & Langlé, R. (2014). La visión de los actores sociales frente a los proyectos eólicos en el Istmo de Tehuantepec. https://consultaindigenajuchitan.files.wordpress.com/2015/01/l-eolico.pdf

Oaxaca Economy Ministry. (2010). Invitación para el Coloquio 2010. https://www.amdee.org/Coloquio_Oaxaca/invitacion.pdf

Oceransky, S. (2008). *Wind conflicts in the isthmus of Tehuantepec: The role of ownership and decision-making models in indigenous resistance to wind projects in Southern Mexico.* Proceedings of the 7th world wind energy conference 2008: Community power: Energy autonomy for local economies, Kingston, Canada. https://www.osti.gov/etdeweb/biblio/21095642

ProjectCo. (2019). *Environmental and social impact assessment for the proposed 60 MW solar power plant in Kanzimbe Village, TA Kalonga, Salima District, Central Region, Malawi.* Final Report. https://jcmpower.ca/wp-content/uploads/2019/03/JCM-Malawi_ESIA_Final-Report_V6_March-2019.pdf

Sovacool, B. K., & Brisbois, M.-C. (2019). Elite power in low-carbon transitions: A critical and interdisciplinary review. *Energy Research & Social Science, 57*, 101242. https://doi.org/10.1016/j.erss.2019.101242

Sovacool, B. K., Burke, M., Baker, L., Kotikalapudi, C. K., & Wlokas, H. (2017). New frontiers and conceptual frameworks for energy justice. *Energy Policy, 105*, 677–691. https://doi.org/10.1016/j.enpol.2017.03.005

Sovacool, B. K., & Dworkin, M. H. (2015). Energy justice: Conceptual insights and practical applications. *Applied Energy, 142*, 435–444. https://doi.org/10.1016/j.apenergy.2015.01.002

Sultana, F. (2007). Reflexivity, positionality and participatory ethics: Negotiating fieldwork dilemmas in international research. *ACME: An International Journal for Critical Geographies, 6*(3), 374–385. https://142.207.145.31/index.php/acme/article/view/786

Tarekegne, B. (2020). Just electrification: Imagining the justice dimensions of energy access and addressing energy poverty. *Energy Research & Social Science, 70*, 101639. https://doi.org/10.1016/j.erss.2020.101639

Taulo, J. L., Gondwe, K. J., & Sebitosi, A. B. (2015). Energy supply in Malawi: Options and issues. *Journal of Energy in Southern Africa, 26*(2), 19. https://doi.org/10.17159/2413-3051/2015/v26i2a2192

Upham, P., Oltra, C., & Boso, À. (2015). Towards a cross-paradigmatic framework of the social acceptance of energy systems. *Energy Research & Social Science, 8*, 100–112. https://doi.org/10.1016/j.erss.2015.05.003

Velasco-Herrejon, P., & Savaresi, A. (2019). Wind energy, benefit-sharing and indigenous peoples: Lessons from the Isthmus of Tehuantepec, Southern Mexico. *SSRN Electronic Journal, 18*,1-16. https://doi.org/10.2139/ssrn.3337142

Wolsink, M. (2007). Planning of renewables schemes: Deliberative and fair decision-making on landscape issues instead of reproachful accusations of non-cooperation. *Energy Policy, 35*(5), 2692–2704. https://doi.org/10.1016/j.enpol.2006.12.002

Wood Group. (2018). *Powering development in Mulanje, end of year 3 report.* https://www.gov.scot/binaries/content/documents/govscot/publications/transparency-data/208/08/malawi-development-programme-2015-2018-year-three-reports/documents/m-15-r-019-powering-development-in-mulanje%2D%2D-year-three-report/m-15-r-019%2D%2D-powering-development-in-mulanje%2D%2D-year-three-report/govscot%3Adocument

Woods, M. (2003). Conflicting environmental visions of the rural: Windfarm development in mid wales. *Sociologia Ruralis, 43*(3), 271–288. https://doi.org/10.1111/1467-9523.00245

World Bank. (2020). *Access to electricity (% of population)—Malawi | Data.* https://data.worldbank.org/indicator/EG.ELC.ACCS.ZS?locations=MW

Wüstenhagen, R., Wolsink, M., & Bürer, M. J. (2007). Social acceptance of renewable energy innovation: An introduction to the concept. *Energy Policy, 35*(5), 2683–2691. https://doi.org/10.1016/j.enpol.2006.12.001

Yenneti, K., & Day, R. (2016). Distributional justice in solar energy implementation in India: The case of Charanka solar park. *Journal of Rural Studies, 46*, 35–46. https://doi.org/10.1016/j.jrurstud.2016.05.009

Zárate Toledo, E., & Fraga, J. (2019). El derecho de consulta previa en la transición energética mexicana. *Cahiers Des Amériques Latines, 90*, 123–140. https://doi.org/10.4000/cal.9191

PART VII

Discussion

CHAPTER 13

Contributions, Tensions and Future Avenues of a Critical Approach to the Social Acceptance of Renewable Energy Infrastructures

Susana Batel and David Rudolph

Introduction

The main aim that we had envisaged for this book was to bring together scholars from the energy research and social sciences field who have been specifically reflecting upon the social acceptance of renewable energy technologies (RET) in the last decades, to contribute to build and consolidate a critical approach to research in this area. As highlighted in the Introduction, all chapters contributed to that main aim by accepting our call to engage with that critical approach and particularly by illustrating one of its main tenets or dimensions as proposed by us: *overcoming*

S. Batel (✉)
Instituto Universitário de Lisboa (ISCTE-IUL), Cis-IUL, Lisboa, Portugal
e-mail: susana.batel@iscte-iul.pt

D. Rudolph
Department of Wind Energy, Technical University of Denmark, Roskilde, Denmark
e-mail: dpru@dtu.dk

S. Batel, D. Rudolph (eds.), *A critical approach to the social acceptance of renewable energy infrastructures*,
https://doi.org/10.1007/978-3-030-73699-6_13

individualism and socio-cognitive reductionism, *repoliticisations*—recognising and articulating power relations, *interventions*—praxis and political engagement with research, *for interdisciplinarity* and *overcoming localism and spatial determinism*. These dimensions mirror as much the cornerstones that a critical research approach entails (e.g. Batel & Adams, 2016), as the initial contemplations by us over potential gaps in social acceptance of RET research that are indicative of the need for more critical assessments of the directions that social acceptance research has taken.

Contributions from the Chapters

In addition to our initial overall framing, the chapters that were presented also often better exemplified, expanded and/or opened up new lines of inquiry to the critical approach to the social acceptance of RET that this book proposed. This does not mean that our initial conceptualisation has become obsolete. Instead, the diversity of chapters has demonstrated that there is no uniform appraisal of the needs for critical research, and where a critical approach should set in, what it should address and what it can accomplish. While all the chapters speak to the original themes and are arranged along the original layout, they address various topics and their thematic, ontological and epistemological contributions vary considerably. These contributions can be boiled down and tentatively organised around four main aspects.

First, the need to further focus on the social acceptance of already operational RET projects. A repoliticisation, better understanding and exploration of what happens at this stage of RET deployment are crucial for fostering more just transitions, as argued by Rudolph and Clausen and also well illustrated by Dan van der Horst and colleagues with their analysis of social acceptance and energy justice in different socio-spatial and cultural contexts of the Global South, namely, wind farms in Oaxaca, Mexico, and hydro mini-grid in Mulanje, Malawi. As van der Horst and colleagues put it, research on the social acceptance of RET, both in the Global South and in the Global North, has to stay more "attuned to conditions of social acceptance, whilst these are diverse and non-static throughout the lifecycle of RET" (Chapter 12, p. 104), which significantly includes the post-construction phase. Paying more attention to this phase of the lifecycle of RET becomes even more relevant as time goes by, bringing with it socio-political, economic and technological changes that create different alternatives and imaginaries of what a low-carbon energy transition can mean. More decentralised models of renewable energy generation and consumption, at household and community levels, are nowadays more available. As they get disseminated in adverts and the media and get territorial presence, they confront communities with potential alternatives to the business-as-usual, centralised large-scale model of RET deployment, which arguably generates much more adverse impacts and accounts for more unfair decision-making processes (but see van Veelen, 2018).

A second strand of chapters in this book calls our attention precisely to the need for conducting research on the social acceptance of RET that pays more attention to the socio-historical contingencies and complexities of the contexts in which RET are being deployed. The contributions by Sherren and colleagues, Batel and Devine-Wright, Gismondi and Hanson, van der Horst and colleagues and Improta and Di Masso illustrate in a clear way how energy, spatial and land use pasts and associated power relations, social practices and materialities are intricately associated with people's current responses to energy transitions. These chapters illustrate well how the lived experiences of RET and of other energy sources, infrastructures and relations, shape people's engagements and practices with RET and have therefore to be more taken into account in analyses of the social acceptance of RET.

Third, this brings us to another clear line of inquiry that is expanded and brought to the fore with some of the contributions presented in this book—on considering more the role of other actors, components and entities in shaping the social acceptance of RET. These can be objects, planning procedures, permits, birds and devices to measure wind, as proposed by Kirkegaard and Nyborg as well as Nadaï and Labussière; and other actors besides local communities, such as developers, policy-makers and other stakeholders involved in energy transitions, as discussed by Cuppen and Pesch. These authors, as well as the chapter by Rudolph and Clausen, show through very concrete and clear examples, how our activities as researchers, the questions we ask and the questions we let other actors—humans and non-humans—ask us, can impact on this field of research and on energy transitions themselves. This is also discussed and illustrated, albeit in different ways, in the contribution by Haggett—from a more discursive, socio-constructionist perspective—and by Nadaï and Labussière as well as Kirkegaard and Nyborg—based on more material, experimental ontology perspectives.

Finally, a fourth line of inquiry that clearly is also taken to new directions by some of the contributions in this book is on further analysing the social acceptance of RET as a reflection, cause and consequence of the larger socio-economic and political context in which it is embedded and through which it is shaped, namely, neoliberal capitalist societies. Dunlap's chapter is most paradigmatic of this line of inquiry, going as far as proposing even the incommensurability of RET, as it were, with any pretence to environmental sustainability, green democracy, or post-carbon transitions. RET should only, at the most, be conceived as fossil fuel+, given that in all the instances of their life cycles they originate from and materialise neoliberal capitalism. The chapters by Batel and Devine-Wright as well as Improta and Di Masso also point to this relationality between people's responses to RET and the larger socio-economic and political contexts in which they are articulated and through which they take shape. Batel and Devine-Wright argue that the relation between the visual-spatial impact of RET and their social acceptance is not an individual process, but instead reflects a pervasive interplay between positional, community and cultural/ideological dimensions. In turn, Improta and Di Masso show that the social acceptance of RET in the Global South is still permeated by one-sided structural power relations between the Global North and the Global South, now in the guise of large-scale wind farms and the promise of development and progress presented as a panacea offering to economically deprived communities.

In sum, the reflection and discussion that the chapters offer of the main axes that we proposed to define a critical approach to the social acceptance of RET seem to suggest that such a critical approach is much-needed in order to further conceptualise, analyse and intervene more deeply in the repoliticisation of social acceptance research and recognising and articulating inherent power relations, namely, by identifying and examining the socio-spatial, economic, political, ideological, cultural and historical contingencies of social acceptance. In fact, and even if each chapter aimed at particularly illustrating one of the tenets of the critical approach to the social acceptance of RET that we aimed to present and develop in this book, something that the chapters clearly and equally carry is that understanding social acceptance of RET through a critical approach involves addressing and considering all its tenets and dimensions instead of merely singling one of them out.

Ontological, Methodological and Normative Tensions of a Critical Approach

Despite the commonalities across the chapters, the contributions also interestingly highlight that what critical might mean and how it should be practiced within social acceptance research is not univocal and actually entails different interpretations and approaches applied to different extents in the authors' research. Two key ontologies that seem to translate a critical approach to social acceptance research across the chapters are relationality and materiality. Relationality is taken up as a constitutive part of social acceptance in different ways: for instance, for Sherren and colleagues, "There is a relational turn underway in human-environment research that locates values or subjective states such as resilience as emergent between people or between people and environments, a product of their interactions, rather than residing with any one of them *per se* [...] This relational turn is an invitation in our survey methods to look past the respondent and include social variables that indicate how the respondent is socially located, how their livelihood is sustained, and the levels of trust that exist between that individual and a range of relevant institutions and organisations" (p. 27, Chapter 2). In turn, Nadaï and Labussière state they have used throughout their research "what we could name relational approaches—i.e. approaches that do not take individualities or entities as given and stable, but rather as in-the-making through processes and relations. This meant to follow, describe and acknowledge the way in which processes of developing renewable energies embarked entities—humans and non-humans—and changed ways of relating with / among them" (p. 144, Chapter 8). It seems then that this emphasis on relationality and the embracing of 'flatter' ontologies (Ash, 2020) to the detriment of more structuralist ones, is also a core aspect of a critical approach to the social acceptance of RET (see also Batel, 2018; Labussière & Nadaï, 2018).

Materiality is also a key focus of most of the contributions even if being defined and approached in diverse ways. These seem to reflect the tension between 'old' and 'new' materialisms, between Marxian historical materialism (as in Dunlap's chapter; see also Bellamy & Diamanti, 2018) and the materialism of human actors and of objects or actants within Actor-Network-Theory assemblages (well-illustrated by Kirkegaard and Nyborg); between the material effects of discourses (as in Haggett's chapter) and the consequences of existing infrastructure for local communities' livelihoods (as suggested by Sherren and colleagues) and for emerging actors (as in the chapter by Gismondi and Hanson), and the material participation of all entities involved in energy transitions (as discussed by Nadaï and Labussière). Relationality and materiality are then closely intertwined as ontological dimensions of a critical approach to the social acceptance of RET (see also Batel, 2018; Labussière & Nadaï, 2018), but their different translations in this book, especially regarding materiality, highlight that one still existent tension within this area is the relation between the role of the social and of the material or technical in social acceptance and whether we can consider them to embrace the same ontological level. Nadaï and Labussiére suggest that we, as social scientists, should "reflexively engage with the field and be challenged by it, because this paves the way to enlarged experimentation and ontologies, and thus new possibilities" (p. 152, Chapter 8). While this is a key insight feeding into research that has been increasingly highlighting the need for more bottom-up ontologies—such as research contesting green energy colonialism (e.g. Normann, 2020) and research adopting a capabilities approach to better understand local communities' responses to RET (e.g. Velasco-Herrejon & Bauwens, 2020)—it is also crucial to acknowledge that we can never completely suspend how the relations between humans and non-human entities (and including ourselves as social scientists) are currently organised, and their positionings and materialisations within the larger socio-economic, political and ideological systems we live in. This has also been recently pointed out by Ash (2020), in relation to flat ontologies, that still often seem to fall short of explicitly recognising, articulating and proposing ways to address power relations, and particularly structural ones. In other words, being critical not only is to shed light on the politics and consequences of RET but implies explicitly adopting a particular normative perspective that assumes and aims to uncover and contest existent power relations and any related forms of oppression and injustice (see also Batel, 2020), be it between developers and local communities, social scientists and the communities they study, or within local communities themselves and their human and non-human constituents. The essay presented in this book by Dunlap and the empirical analysis of the deployment of a wind farm in Brazil by a Spanish developer presented by Improta and Di Masso are useful examples of that and are indicative of what a more structuralist, dialectic and historical-materialist approach to the social acceptance of RET can contribute to flat ontologies and more experimental relational and material approaches (see also Bellamy & Diamanti, 2018). When submerging relations of difference, power and domination into a flat and symmetric conglomerate, there is a limit to the extent to which the historical and causal underpinnings of structural differences between human and non-human forms of agency as well as relations of power between humans can be revealed. Indeed, a differentiation of entities would need to be based on capacity, degree and form, rather than being reduced to a certain type or kind (Ash, 2020). But, only by retaining an *historical* agential distinction between different material and social entities is it possible to acknowledge asymmetric power relations and understand how various agents have acquired their powers to act and how their agency has developed and becomes enacted in certain processes (Choat, 2018). A key illustration lies in the fundamental condition of capital accumulation being the constant incorporation of external material entities in capitalist cycles of production through processes of coding, quantification, commodification and marketisation to serve economic growth and generate uneven development (Smith, 1984).

This type of materiality and relationality presumed under the critical approach we aimed at developing in this book should then help uncover and address the neoliberal and capitalist, and thus exploitative, oppressive and uneven entanglements of RET. As such, we can say that in this book and with the help of the rich arguments and illustrations of the chapters that compose it, we provocatively propose that a critical approach to the social acceptance of RET would need multiple ontologies to entail an intertwined structuralist *and* flat relational perspective at the ontological level in order to exploit its full epistemological potential. This also entails an understanding of materiality that takes account of both structural power relations and the networks and assemblages of human and non-human constituents that materialise them. In such a way, the analytical strengths of relational perspectives can be combined with the normative and interventionist principles of structural perspectives in a critical approach towards the contested character of acceptance and change, without disregarding one or the other (e.g. Bridge & Gailing, 2020). Some of these challenges may be alleviated by recognising and acknowledging "the difference between drawing boundaries in an analytical and an ontological sense" (Hornborg, 2014, p. 127). Hence, interdisciplinarity not only necessitates a greater awareness of terminological divergences, as strikingly demonstrated by Haggett's contribution, but also requires us to carefully address ontological discrepancies between disciplines and approaches to the benefit of radical change. This articulation of multiple ontologies also allows, as proposed by Gismondi and Hanson, to "avoid a simple David and Goliath interpretation" (p. 180, Chapter 10) of RET in relation to neoliberal capitalist systems, and account for how other power relations, at different scales, can be constraining and generative both of 'socially acceptable' or consented (Webler & Tuler, 2020) ways to deploy RET and of radical, transformative, bottom-up and experimental alternatives to RET.

This discussion on the definition and role of relationality and materiality within a critical approach to the social acceptance of RET is also relevant, because it also reflects how, in the move towards overcoming the individualism, socio-cognitivism and quantification that permeated research on the social acceptance of RET over several decades (Ellis et al., 2007; Normann, 2020), the lived and experiential dimensions of people's relations with RET have become quite forgotten and are almost considered as unimportant. Hence, Dunlap points out, research so far is "still qualitatively limited in documenting the profound psycho-social and ontological impacts of energy extractivism" (p. 94, Chapter 5). This has also been illustrated by Improta and Di Masso, with their analysis on how place-talk and arguments construct specific person-place psychological bonds that contribute to the perpetuation and legitimisation of energy colonialism (see also Batel, 2018; Dunlap & Sullivan, 2020; Groves, 2015; Normann, 2020). It is then essential that in our strive for interdisciplinarity within a critical approach to the social acceptance of RET, we do not dismiss the imports of disciplines such as social, environmental and community psychology and anthropology—or at least of some strands within those disciplines -, given that a crucial part of a critical approach to the social acceptance of RET needs to explore and understand how people and communities live with, experience and relate to RET in their everyday lives, and what impacts that may have, at symbolic, psycho-social, subjective and community levels—including the embodied injustices from the supply chains of RET (Healy et al., 2019).

This brings us to what has been another key tension or debate within research on the social acceptance of RET; the tension between the use of different methods, associated ontologies and epistemologies to understand the social acceptance of RET. Ellis and colleagues, in their influential 2007 paper, relevantly highlighted how research on the social acceptance of RET up until then tended to be only focused on the local and adopt mainly individualist, positivist and quantitative approaches to analyse local responses. More than ten years later, other authors continue to highlight the relevance of using more qualitative methods in social acceptance research, and the chapter by Sherren and colleagues originates precisely from this area of debate within research on the social acceptance of RET to propose how quantitative methods can also be used without perpetuating the pitfalls of more positivist and individualist agendas to which they have been often coupled. Interestingly enough, back in 2007, Ellis and colleagues had suggested that Q-methodology could be a promising and useful alternative for research in a more contextualised way to explore people's responses to and relations with RET. In this book, both Sherren et al. and Cuppen and Pesch also make reference to Q-methodology as a method with great potential to provide more nuanced, in-depth and contextualised insights about people's—communities, developers, policymakers—ideas, imaginaries and practices with and about RET.

Eventually, a critical approach also provokes normative tensions with regard to what social acceptance research does and is meant to achieve. Overall, the chapters in the book reveal that the field of research on the social acceptance of RET, while having changed in the last years, has also significantly stayed the same—and the still numerous references to NIMBY and local opposition throughout the chapters, even if to criticise those concepts and foci of research, testify to that. Hence the need and timeliness for proposing and developing with this book a critical approach to the social acceptance of RET that ultimately tries to break with some of the business as usual, depoliticised and fossilised tendencies of current research on the social acceptance of RET and on energy transitions in general. As such, this takes us to reflect again on whether it is still appropriate to refer to the term 'social acceptance' at all, as it presumes that people are the 'hitch', what is to blame, the 'social' as the troubling factor coming from the outside to obstruct and delay the so-called low-carbon energy transitions. In turn, this ultimately reduces the process of energy transition to a techno-scientific task, whereas the purpose of social acceptance research remains confined within the scope of generating acceptance of the formative technical and economic measures of energy transitions. This binary view then tends to include a hierarchical relationship between the technical and social constituents of energy systems, where the social becomes subordinate to the technical and economic realms, which is also reflected in the misleading obsession with the term 'public acceptance', as recently criticised by Wolsink (2019). In that regard, the term 'social acceptance' seems to altogether obscure the relationality and materiality of the technical, historical, economic, political, spatial and not only social systems in which the social acceptance of RET are embedded and generated. Some authors have suggested that it might be useful to use this term as a reference to this specific area of research (Dermont et al., 2017; Ellis & Ferraro, 2016), namely in connection to the research agenda set out by Wüstenhagen, Bürer and Wolsink in 2007. Yet, it might be relevant to further think about what is actually gained with maintaining this concept or with abandoning it altogether and rather focusing on socio-technical transitions and responses to those at different scales and levels. In fact, one of the key insights of the social acceptance triangle of renewable energy innovation as proposed by Wüstenhagen et al. (2007) was their intention to make explicit that social acceptance is based on three different but *interrelated* dimensions—socio-political, community and market acceptance—all of which operate at different scales and in relation to distinct actors, and are shaped by different socio-psychological, political, economic and cultural factors. On the one hand, while this was clearly relevant and impactful at the time to precisely bring more to the fore the relevance of the social and considering people's responses at different levels, this model is also locked-in in a very specific way of 'doing' renewable energy transitions, in more market-driven, centralised and growth-driven principles. On the other hand, it is precisely the proposed interrelatedness of the three dimensions that would allow for a more critical perspective on the issue of social acceptance of RET, but which has rarely been employed as such and has become mashed up in public acceptance as an externalised proxy of social acceptance (Wolsink, 2019). Thus, the concept also seems to leave aside the very materiality—if not relationality (see Batel, 2018)—of the social acceptance of RET that we have discussed here, and that must clearly include non-human constituents and relations as much as human ones.

Questioning the very use of the concept of social acceptance to discuss people's relations and practices with RET might also be useful for this area of research to more easily engage with the whole life cycle of RET and the groups, scales and impacts involved throughout (Healy et al., 2019). Important social issues such as global energy and environmental justice in relation to RET, how Global North and Global South relations and associated neo-colonialisms are reproduced or contested by RET-related decisions, processes and infrastructures, have been rather neglected by research on the social acceptance of RET so far (Batel & Devine-Wright, 2017; Healy et al., 2019; Normann, 2020; also Improta & Di Masso this book). The fact that this body of research and social acceptance tend to be defined by acceptance in the Global North is quite evident, even in this book, where most contributions are from authors in the Global North who write about the Global North. So there again, a move away from the concept of social acceptance might arguably help with these issues to become core issues to be taken into account, analysed and addressed within this area of research. The import of an energy justice approach, such as proposed by van der Horst and colleagues, might be useful for that, and especially, as recently suggested and illustrated by Velasco-Herrejon and Bauwens (2020), if combined with a bottom-up approach, focused on a definition of energy justice and associated issues by local, indigenous communities themselves. This is also similarly argued by Rudolph and Clausen who highlight the value of place-based and bottom-up approaches to RET that are "scaled to local needs, values and collective benefit-sharing" (p. 74, Chapter 4).

In fact, the relevance of adopting such a bottom-up, contextual approach, mindful of the situated knowledge of local communities and hence of the socio-cultural, geographical, historical and material specificities that shape the appropriation, negotiation and contestation of energy transitions in different groups and places is something that this book also clearly illustrates and highlights as crucial for a critical approach to the social acceptance of RET. The insights from different case studies in Brazil, Canada, Denmark, France, Malawi, Mexico, the Netherlands and the UK that were presented also open up the discussion on whether being critical implies not only overcoming localism and spatial determinism, as we have suggested as a main premise of a critical approach to the social acceptance of RET, but also whether a critical approach has to be contextualised as well? This becomes particularly relevant when considering the uneven development stages of RET deployment in different countries. Is being critical in a certain context, such as in relation to the energy system in Portugal, different from being critical in Canada or in Malawi? Dunlap's reflection on RET as fossil fuel+ contributes to this discussion by highlighting the relationality of energy, capital and development regimes within current socio-economic and political systems. This is also well illustrated by van der Horst and colleagues, suggesting that to be critical in some contexts might actually entail promoting what a critical approach criticises in other contexts, such as certain types of socio-economic development and infrastructures. Being critical would then also ultimately mean to turn away from a one-sided view of RET rationales as a substitute of fossil fuels to maintain economic growth and to focus more on what RET may prompt or may be able to achieve in terms of social transformation and justice at various scales and different contexts. In consequence, the rationale of critical social acceptance research would need to become more liberated from the underlying technocratic approach of replacing fossil fuels with renewables and increasing RET capacities, and pay more attention to intended, unintended and hidden consequences of this substitution as well as the political, economic and spatial processes that give rise to them. Social acceptance of RET would then need to be engaged more with the broader socio-economic context, which not only shapes responses to RET development, but that also engenders a myriad of uneven consequences and outcomes.

Future Avenues for Critical Social Acceptance Research

The insightful reflections, re-conceptualisations and discussions offered so far throughout this book, ask for some final thoughts on where they could and should take us when we consider future avenues for critical social acceptance research. We propose that these avenues can be tentatively organised along two constellations: On the one hand, avenues for research that follow more radical, epistemological and political shifts in relation to current research; and on the other hand, avenues for future research that further push for specific lines of inquiry of a critical approach that are already, to some extent, underway, some of which discussed earlier in this book.

Beginning with the former constellation of avenues for future research based on a critical approach, a deeper focus on the political is the evident way forward, with the political taken as "the study of power and domination [...] including the gendered and racialized politics" (Daggett, 2020, p. 17) of energy eco-socio-technical systems, including those involving renewables. This is a critical uptake that has been largely lacking in both more representational and more materialist approaches to the social acceptance of RET (Batel, 2020) and that implies more explicitly acknowledging and analysing the embeddedness of RET within (neoliberal) capitalism, its structural inequalities and intersectional injustices and associated circuits of privilege and dispossession (Kaijser & Kronsell, 2014; Malm, 2016). The entanglements between (renewable) energy and related extractive industries; the ecological crisis; gender inequalities; poverty and migration; racism, colonialism and other forms of oppression; within current systems, need to be made much more explicit and brought to the fore

(Baka, 2017; Bell et al., 2020; Daggett, 2020; Lennon, 2017; Listo, 2018). Likewise, the historical entanglements of these energy and socio-ecological struggles and associated past and present contentious politics need to be explored more clearly (Paul, 2018). In turn, such an ontological and political shift into the relationality, pervasiveness and one-sidedness of traditional capitalist relations of domination necessarily also implies adopting more relational and holistic perspectives into renewable energy systems themselves as technical configurations encompassing, as it were, the extraction of materials for their construction, being connected to power lines, storage systems, the consumption of electricity and so on and so forth. To this end, research already conducted in the Global South might provide important insights, as it has already more clearly explored and examined energy systems across their life cycles, often within a distributive justice perspective, as illustrated in the chapters by van der Horst and colleagues and also by Improta and Di Masso. Another related key avenue for future critical social acceptance research has to venture beyond being critical within the system, coupling it with understanding how to actually "overcome the deep roots of capitalism's ever-growing energy dependence" (Bellamy & Diamanti, 2018; p. x; Daggett, 2020). As already hinted at before, the fundamental idea of renewables is to decouple growth from fossil fuels, but to what extent is that possible? (Brand & Wissen, 2018; Siamanta, 2019). This avenue should then consider how a desired growth of RET achieves a transition away from fossil fuels rather than adding to and expanding the amount of energy production (York & Bell, 2019). An existential shift needs to happen as well on social acceptance research for it to accommodate acceptance of degrowth as part of the expansion of renewable energy infrastructures and associated eco-socio-technical systems. All these considerations ultimately highlight the role of repoliticisations but also hint at a tension among some chapters that seems to implicitly resonate with the epistemological variances related to the very term of 'repoliticisation', between 'making things public' by tracing the processes of how certain things and issues become 'matters of concern' (e.g. Krauss, 2010) and the purposeful renegotiation and deliberation of depoliticised issues with the aim of not only exposing power relations but also shifting the powerful frames of an issue in order to bring about change (Bues & Gailing, 2016), which should also be taken up more thoroughly in the future.

Besides these avenues for future research that encompass more profound epistemological and political shifts, there are also certainly more particular lines of inquiry and future research agendas that should be more consistently pursued, explored and developed, specifically in relation to three main areas: the increasing financialisation of RET and the public versus private divide; energy democracy and public participation in energy decision-making; and the co-creation, transferability and impact of social sciences' energy research to energy policy-making, political agendas, local communities and other affected actors.

An increasingly important but so far less explored facet of the RET development relates to the growing influence of financial capital and financial institutions in the green economy that likewise affect the deployment, technological orientation and diffusion of RET (Klagge & Nweke-Eze, 2020; Knuth, 2018). Financialisation and large investments in RET projects that are predominantly driven by profit maximisation and private shareholder interests open up a focus of research that not only brings together and affects the socio-political, community and market tenets of social acceptance, but also raises important questions about distributional justice, energy colonialism and innovation trajectories. This also leads us to the alleged binary between public and individual or private interests that has been famously described to form the 'social gap' (Bell et al., 2013) between large societal acceptance of RET and local resistance of particular RET projects. However, this division does not do fully justice to the varied melange of interests and can hardly be sustained. Instead, there are various conflicting goals between action-guiding goals of energy transition and goals articulated as public interests. Likewise, private interests related to RET are not monolithically bound to members of local communities either, but can also comprise the interests and goals of incumbent and emergent energy companies, shareholders or community energy collectives (Krüger, 2020). Hence, this public-private binary appears as too static and does not sufficiently reflect the dynamic nature of contestations over RET and their contextual embeddedness. Critical research on social acceptance should rather be dedicated to the changing relations between interests related to the common good vis-a-vis the particularism of vested interests, especially with regard to how neoliberal political ideologies of privatisation, responsibilisation and enclosure encroach upon collective interests and interact with issues such as social fairness, democracy, equality, living economy and rural development (e.g. Furnano, 2019; Gudmundsdottir et al., 2018; Hess, 2012; Siamanta, 2019). This research avenue would particularly involve a critical assessment of repercussions for distributional justice between community-driven and large-scale RET projects that have tended to drift apart as two different areas of research, and the exploration of new possibilities for impeding this trend.

Also concerns about the lack of legitimacy of renewable-energy-related decision-making processes and their impacts and consequences for different communities have been at the heart of the recent development of research agendas on energy democracy and energy justice (Becker & Naumann, 2017; Feldpausch-Parker et al., 2019; Jenkins et al., 2016; Sovacool & Dworkin, 2015; van Veelen & van der Horst, 2018). This seems to be another symptom of the inaptitude of democratic systems that have been designed and materialised alongside neoliberal capitalism to properly allow for taking society on board in energy and environment decision-making while reproducing an inherent separation of the social from the technical aspects of energy transitions. In other words, the key challenge going forward is not to improve the participation process in the deployment of specific RET but instead to fundamentally reconsider the role of the public and of participation in energy decision-making and within democratic systems altogether (Batel, 2018; Cuppen, 2018, for some discussions). In a related way, it is crucial that any conceptualisations and research on energy democracy and associated ideas of empowerment and participation are developed within a critical light that does justice to what they should imply and that does not just co-opt them as another expression or consequence of a neoliberal agenda. This includes fundamental questions regarding the rationales, foundations and manifestations of empowerment and participation vis-à-vis those related to responsibilisation, governmentality or asset-based development. In turn, this critical look at participation and empowerment in relation to RET deployment and related decision-making processes also implies, as already argued, an exploration of the manifestation and consequences of RET after their deployment and the options and opportunities for people's engagement with, negotiation of and resistance to infrastructures and decisions to which they might not have given consent (Webler & Tuler, 2020).

The continued expansion of renewable energy has resulted, at least in some countries, in a condition where RET have already become critical infrastructures forming substantial and formative cornerstones of the energy system. Since RET have grown out of their infancy, or are in the process of growing out, and have evolved to critical infrastructures, it is necessary to diverge away from the narrow rationale of acceptance or opposition being the bottleneck of RET development. While this is certainly still a resurgent issue when it comes to the local deployment of specific RET projects, problematisations would need to be reframed and research would have to inquire more broadly into what this maturing condition of RET may then mean for the 'social acceptance of renewable energy innovation' (Wüstenhagen et al., 2007) and what goals this innovation and its diffusion may address, include and target beyond the growth of energy capacities and geographical expansion. From a critical perspective, this should include questions as to how expanding RET can help alleviate or revert the structural inequalities and intersectional injustices produced in the evolution and formation of a carbon-fuelled energy system and contribute to socio-ecological transformations leading to more just and equal post-carbon societies.

Finally, another key pathway for future research adopting a critical approach to social acceptance is to adopt more non-exclusive ontological considerations and, as suggested by Haggett, consider more and in a more analytical way the role of terminological clarifications in energy social science research, such as in relation to current fetishisms (Cunha, 2015), like community, empowerment and materiality and related assemblages. This body of research already abounds with examples of how concepts are not just words, especially when they become reified and fetishised as explanans of central social issues, such as the NIMBY label (Burningham, 2000; Devine-Wright, 2005) or the very concept of social acceptance (Batel et al., 2013). While changing the name of something to something else—such as social acceptance (Nadaï & Labussière, Chapter 8, p. 142), or the Anthropocene to the Capitalocene, as discussed by Cunha (2015)—will not "solve the underlying social and material contradictions" (Cunha, 2015, p. 67) of a capitalist fossil fuel (or renewables …) based system, it certainly allows us to create the required reflective space to contemplate over what those social and material contradictions entail for being able to devise ways to overcome them and to create fairer and more just eco-socio-technical energy futures. As put clearly by Gismondi and Hanson (p. 180, Chapter 10) "struggles over meaning-making and re-presentation are actions, not just word games", especially given that we, as researchers, are "part of the rhizome, and as such have to account for the conceptions and categorisations that we introduce" (Cuppen & Pesch, p. 169, Chapter 9; see also Haggett, Chapter 7).

Eventually, and regardless of the directions of the future avenues for critical research, there is an urgent need for a continued and stronger transfer of research findings into different realms of energy transition practice. However, an unsurprising challenge consists in the obvious discrepancy between the emancipatory intentions of critical research and the hegemonic business-as-usual approach within depoliticised energy transition practice, which generally tends to decouple theory and practice. Addressing this challenge becomes important in at least three regards. First, critical research should not stop at tracing and describing societal developments and their implications, but also intervene in existing power relations, historically grown injustices and socio-political contestations. Second, critical research would have to formulate research questions that are relevant and appealing to stakeholders, practitioners and social movements that strive for transformational and emancipatory processes. Third, critical research on the social acceptance of RET could team up and start a more profound dialogue with neighbouring disciplines in order to bring together critical and applied perspectives and to have a stronger impact. All this necessitates scholarly activism, and engagement and collaboration within and outside academia with stakeholders from the private energy sector, public sector and civil society, and while considering non-human actors as well.

Conclusions

The continuous implementation, expansion and socio-political establishment of RET should not short-sightedly (nor in an extricating manner) be considered as a sign of social acceptance research becoming obsolete, instead it is more an indication that the auspices of this research area have changed. This calls for and ultimately necessitates a realignment and re-imagination of the ontological and epistemological premises of social acceptance research. The primary objective should no longer involve a problematisation of the 'social' dimensions in order to find 'acceptable' solutions to pave the way for technological transformations, but instead embody a critical engagement with the development of RET as social and political projects capable of holistic transformations that venture beyond the political-economic status quo and fossilised understandings of growth, development and well-being. This can only be achieved by putting social acceptance research on a more critical footing that turns the focus towards the contextual and consequential relations of expanding RET as a crucial component of energy and societal transitions.

References

Ash, J. (2020). Flat ontology and geography. *Dialogues in Human Geography*. https://doi.org/10.1177/2043820620940052

Baka, J. (2017). Making space for energy: Wasteland development, enclosures, and energy dispossessions. *Antipode, 49*(4), 977–996.

Batel, S. (2018). A critical discussion of research on the social acceptance of renewable energy generation and associated infrastructures and an agenda for the future. *Journal of Environmental Policy & Planning, 20*(3), 356–369.

Batel, S. (2020). Research on the social acceptance of renewable energy technologies: Past, present and future. *Energy Research & Social Science, 68*, 101544.

Batel, S., & Adams, M. (2016). Ecological crisis, sustainability and social worlds: Developing a critical agenda. *Papers on Social Representations, 25*(1), 1.1–1.27.

Batel, S., & Devine-Wright, P. (2017). Energy colonialism and the role of the global in local responses to new energy infrastructures in the UK: A critical and exploratory empirical analysis. *Antipode, 49*(1), 3–22.

Batel, S., Devine-Wright, P., & Tangeland, T. (2013). Social acceptance of low carbon energy and associated infrastructures: A critical discussion. *Energy Policy, 58*, 1–5.

Becker, S., & Naumann, M. (2017). Energy democracy: Mapping the debate on energy alternatives. *Geography Compass, 11*(8), e12321.

Bell, D., Gray, T., Haggett, C., & Swaffield, J. (2013). Re-visiting the 'social gap': Public opinion and relations of power in the local politics of wind energy. *Environmental Politics, 22*(1), 115–135.

Bell, S. E., Daggett, C., & Labuski, C. (2020). Toward feminist energy systems: Why adding women and solar panels is not enough. *Energy Research & Social Science, 68*, 101557.

Bellamy, B. R., & Diamanti, J. (Eds.). (2018). *Materialism and the critique of energy*. MCM.

Brand, U., & Wissen, M. (2018). *The limits to capitalist nature: Theorizing and overcoming the imperial mode of living*. Rowman & Littlefield.

Bridge, G., & Gailing, L. (2020). New energy spaces: Towards a geographical political economy of energy transition. *Environment and Planning A: Economy and Space, 52*(6), 1037–1050.

Bues, A., & Gailing, L. (2016). Energy transitions and power: Between governmentality and depoliticization. In L. Gailing & T. Moss (Eds.), *Conceptualizing Germany's energy transition: Institutions, materiality, power, space* (pp. 69–91). Palgrave-Pivot.

Burningham, K. (2000). Using the language of NIMBY: A topic for research, not an activity for researchers. *Local Environment, 5*(1), 55–67.

Choat, S. (2018). Science, agency and ontology: A historical-materialist response to new materialism. *Political Studies, 66*(4), 1027–1042.

Cunha, D. (2015). The Anthropocene as fetishism. *Mediations, 28*(2), 65–77.

Cuppen, E. (2018). The value of social conflicts. Critiquing invited participation in energy projects. *Energy Research & Social Science, 38*, 28–32.

Daggett, C. (2020). Energy and domination: Contesting the fossil myth of fuel expansion. *Environmental Politics, 30*(4), 644–662.

Dermont, C., Ingold, K., Kammermann, L., & Stadelmann-Steffen, I. (2017). Bringing the policy making perspective in: A political science approach to social acceptance. *Energy Policy, 108*, 359–368.

Devine-Wright, P. (2005). Beyond NIMBYism: Towards an integrated framework for understanding public perceptions of wind energy. *Wind Energy: An International Journal for Progress and Applications in Wind Power Conversion Technology, 8*(2), 125–139.

Dunlap, A., & Sullivan, S. (2020). A faultline in neoliberal environmental governance scholarship? Or, why accumulation-by-alienation matters. *Environment and Planning E: Nature and Space, 3*(2), 552–579.

Ellis, G., Barry, J., & Robinson, C. (2007). Many ways to say 'no', different ways to say 'yes': Applying Q-methodology to understand public acceptance of wind farm proposals. *Journal of Environmental Planning and Management, 50*(4), 517–551.

Ellis, G., & Ferraro, G. (2016). *The social acceptance of wind energy. Where we stand and the path ahead.* JRC Science for policy Report, European Commission, Brussels.

Feldpausch-Parker, A. M., Endres, D. E., & Peterson, T. (2019). A research agenda for energy democracy. *Frontiers in Communication, 4*, 53.

Furnano, A. (2019). Neoliberal energy transitions: The renewable energy boom in Chilean mining economy. *Environment and Planning E: Nature and Space, 3*(4), 951–975.

Groves, C. (2015). The bomb in my backyard, the serpent in my house: Environmental justice, risk, and the colonisation of attachment. *Environmental Politics, 24*(6), 853–873.

Gudmundsdottir, H., Carton, W., Busch, H., & Ramasar, V. (2018). Modernist dreams and green sagas: The neoliberal politics of Iceland's renewable energy economy. *Environment and Planning E: Nature and Space, 1*(4), 579–601.

Healy, N., Stephens, J. C., & Malin, S. A. (2019). Embodied energy injustices: Unveiling and politicizing the transboundary harms of fossil fuel extractivism and fossil fuel supply chains. *Energy Research & Social Science, 48*, 219–234.

Hess, D. J. (2012). The green transition, neoliberalism, and the technosciences. In L. Pellizoni & M. Ylönen (Eds.), *Neoliberalism and technoscience: Critical assessments* (pp. 209–230). Ashgate.

Hornborg, A. (2014). Technology as fetish: Marx, Latour, and the cultural foundations of Capitalism. *Theory, Culture and Society, 31*(4), 119–140.

Jenkins, K., McCauley, D., Heffron, R., Stephan, H., & Rehner, R. (2016). Energy justice: A conceptual review. *Energy Research & Social Science, 11*, 174–182.

Kaijser, A., & Kronsell, A. (2014). Climate change through the lens of intersectionality. *Environmental Politics, 23*(3), 417–433.

Klagge, B., & Nweke-Eze, C. (2020). Financing large-scale renewable-energy projects in Kenya: Investor types, international connections, and financialization. *Geografiska Annaler: Series B. Human Geography, 102*(1), 61–83.

Knuth, S. (2018). "Breakthroughs" for a green economy? Financialization and clean energy transition. *Energy Research & Social Science, 41*, 220–229.

Krauss, W. (2010). The '*Dingpolitik*' of wind energy in Northern German landscapes: An ethnographic case study. *Landscape Research, 35*(2), 195–208.

Krüger, T. (2020). *Gemeinwohlkonflikte in der Energiewende: Eine radikaldemokratische Perspektive auf Energiekonflikte und die Grenzen der Deliberation.* WZB Discussion Paper No. SP III 2020-602.

Labussière, O., & Nadaï, A. (Eds.). (2018). *Energy transitions: A socio-technical inquiry.* Springer.

Lennon, M. (2017). Decolonizing energy: Black Lives Matter and technoscientific expertise amid solar transitions. *Energy Research & Social Science, 30*, 18–27.

Listo, R. (2018). Gender myths in energy poverty literature: A critical discourse analysis. *Energy Research & Social Science, 38*, 9–18.

Malm, A. (2016). *Fossil capital: The rise of steam power and the roots of global warming.* Verso Books.

Normann, S. (2020). Green colonialism in the Nordic context: Exploring Southern Saami representations of wind energy development. *Journal of Community Psychology.*

Paul, F. C. (2018). History, space and energy (struggle) in the German *Energiewende. Geoforum, 91*, 1–9.

Siamanta, Z. C. (2019). Wind parks in post-crisis Greece: Neoliberalisation vis-à-vis green grabbing. *Environment and Planning E: Nature and Space, 2*(2), 274–303.

Smith, N. (1984). *Uneven development: Nature, capital and the production of space.* Blackwell.

Sovacool, B. K., & Dworkin, M. H. (2015). Energy justice: Conceptual insights and practical applications. *Applied Energy, 142*, 435–444.

Van Veelen, B. (2018). Negotiating energy democracy in practice: Governance processes in community energy projects. *Environmental Politics, 27*(4), 644–665.

Van Veelen, B., & Van Der Horst, D. (2018). What is energy democracy? Connecting social science energy research and political theory. *Energy Research & Social Science, 46*, 19–28.

Velasco-Herrejon, P., & Bauwens, T. (2020). Energy justice from the bottom up: A capability approach to community acceptance of wind energy in Mexico. *Energy Research & Social Science, 70*, 101711.

Webler, T., & Tuler, S. (2020). Unpacking the idea of democratic community consent-based siting for energy infrastructure. *Journal of Risk Research*, 1–16.

Wolsink, M. (2019). Social acceptance, lost objects, and obsession with the 'public'—The pressing need for enhanced conceptual and methodological rigor. *Energy Research & Social Science, 48*, 269–276.

Wüstenhagen, R., Wolsink, M., & Bürer, M. J. (2007). Social acceptance of renewable energy innovation: An introduction to the concept. *Energy Policy, 35*(5), 2683–2691.

York, R., & Bell, S. E. (2019). Energy transitions or additions? Why a transition from fossil fuels requires more than the growth of renewable energy. *Energy Research & Social Science, 51*, 40–43.

Index

S. Batel, D. Rudolph (eds.), *A critical approach to the social acceptance of renewable energy infrastructures*,
https://doi.org/10.1007/978-3-030-73699-6

The manufacturer's authorised representative in the EU is Springer Nature Customer Service Centre GmbH, Europaplatz 3, 69115 Heidelberg, Germany. If you have any concerns regarding our products, please contact ProductSafety@springernature.com

Printed and bound by CPI Group (UK) Ltd, Croydon, CR0 4YY
30/09/2026
02225326-0006